THE
24 HOUR
SOCIETY

THE
24 HOUR
SOCIETY

THE RISKS, COSTS AND CHALLENGES OF A WORLD THAT NEVER STOPS

Martin Moore-Ede

PIATKUS

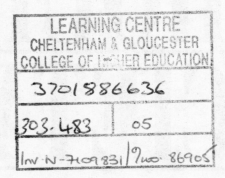
© 1993 Martin Moore-Ede

First published in Great Britain in 1993 by
Judy Piatkus (Publishers) Ltd of
5 Windmill Street, London W1P 1HF

Published in the United States of America by Addison-Wesley.

Many of the designations used by manufacturers and sellers to
distinguish their products are claimed as trademarks. Where
those designations appear in this book, and Addison-Wesley was
aware of a trademark claim, the designations have been printed
in initial capital letters or all capital letters.

**The moral right of the author
has been asserted**

*A catalogue record for this book is available from the
British Library*

ISBN 0 7499 1255 3
0 7499 1269 3 (Pbk)

Text design by David Kelley Design

Printed and bound in Great Britain by
Butler & Tanner Ltd, Frome and London

To my wife,
Dr. Donna Moore-Ede,
in appreciation of her
perceptive wisdom
and loving
support

Contents

PART **II** _____

A SOCIETY IN STRESS

PART **III** _____

THE EMERGING HUMAN
ALERTNESS TECHNOLOGY

PART IV

TRANSFORMATION INTO A HUMAN-CENTERED SOCIETY

Acknowledgments

THIS BOOK tells the interwoven story of two simultaneous twenty-year journeys of exploration, one into the intricate timing mechanisms of the human brain and the other into the around-the-clock operations of industry and government. I could not have accomplished either journey, nor made sense of what I saw along the way, without the education, inspiration, sponsorship, creativity, and guidance generously provided by many people. Unfortunately, as much as I would like to, it is impossible to name them all.

My education has come both from the halls of academia and from the workplaces of industry. My professors at the University of London, Guy's Hospital Medical School, and Harvard Medical School—most notably Professors Jack Hunt, Harry Keen, A. Clifford Barger, John Pappenheimer, James Herd, and Francis D. Moore—taught me how to think, how to rigorously examine and test ideas, and showed me the meaning of excellence in scientific research. My clients, thousands of managers and shiftworkers in over two hundred companies ranging from multinational giants like Exxon, Shell, IBM, General Electric, and Alcan to smaller concerns, took the time out of their busy schedules to educate me into the realities of human performance in around-the-clock operations.

Inspiration has come from those who have encouraged me to turn ideas into reality. Francis D. Moore, Professor of Surgery at Harvard, encouraged me to follow my instincts into unexplored scientific territory; Philip Morrison, Institute Professor of Physics at MIT, showed me how to communicate with clarity and enthusiasm even the most complex ideas; Fred Smith, the Founder and CEO of Federal Express, encouraged me to launch the Institute for Circadian Physiology; and Gary Countryman, Chairman and CEO of Liberty Mutual, showed the value of true friendship, commitment, and generosity when the going got tough.

Sponsorship has come from those who grasped the importance of the mission and were willing to back me with grants, contracts, gifts, and investments from their corporations, foundations, organizations, and governmental agencies. It has taken considerable financial support to develop this new field of scientific research and practical application. Corporate sponsors have included Alcan Aluminum, Alexander and Alexander, Amoco, Atlantic Richfield, Boeing, Boston Edison, Cabot, Caterpillar, Conrail, Dow Chemical, DuPont, Exxon, Foxboro, Liberty Mutual, Madison Gas & Electric, Matshushita Electric Works, Mobil, Monsanto, Raytheon, Southern California Edison, Sun, Texaco, Upjohn, and 3M. Foundation support has come from Amelia Peabody Foundation, Alfred P. Sloan Foundation, Max Kade Foundation, and the Commonwealth Fund. Industry organizations have included Electric Power Research Institute, National Electrical Manufacturers Association, Lighting Research Institute, Motor Vehicle Manufacturers Association, American Bus Association, and United Bus Owners of America. Government agencies providing important research support included the National Institutes of Health, National Science Foundation, NASA, U.S. Air Force, U.S. Army, and the Nuclear Regulatory Commission.

Creative insights and ideas have come from the many fruitful collaborations with my colleagues and students. Frank Sulzman, Charles Fuller, David Kass, David Wexler, Ralph Lydic, Elliott Albers, Margaret Moline, Phillipa Gander, David Borsook, Ziad Boulos, Elizabeth Klerman, and Tom Houpt collaborated in the enquiry into the basic properties of biological clocks. Charles Czeisler, Elliot Weitzman, Richard Kronauer, Gary Richardson, Janet Zimmerman, Scott Campbell, Ted Baker, Drew Dawson, and Claudio Stampi contributed to the exploration of the timing systems in humans. Charles Czeisler, Richard Coleman, Gary Krieger, Jack Connolly, Patricia Click, Faxon Green, Jim Stam, Andrea Sodano, Ted Baker, William Chambers, William Sirois, Martin Stein, and Jan Blais helped develop the applications for industry. William Patrick of Addison-Wesley made sure I communicated these ideas in the most effective way, and Sharon Broll, also of Addison-Wesley, provided additional assistance.

Guidance has come from many who have tried, not always successfully, to keep me on the right track. I particularly appreciate the dedicated support of the Board members of the Institute for Circadian Physiology who have helped nurture this new institution; they include Charles D. Baker, A. Clifford Barger, Gary Countryman, Howard Dyer, Joe Henson, Fisher Howe, Michael Lytton, Francis D. Moore, Richard Pharo, Bernard Reznicek, Robert Seamans, Jr., James St. Clair, and Janet Whitla. I also wish to especially acknowledge the enormous contributions of Susanne Churchill, Birthe Creutz, and Filomena Scibelli.

Finally, this book would not have been written without the vision, understanding, love, and patience of my wife, Dr. Donna Moore-Ede, or the inspiration provided by my children, Andrew and Alexandra. Their encouragement was more important than anything else.

Foreword to the British Edition

WHEN CAPTAIN Tim Lancaster climbed out of his bed on Sunday morning June 10, 1990 after a restful night of sleep, little did he expect what was in store for him. He was scheduled to fly a British Airways BAC 1–11 with 81 passengers down to Malaga that morning. The take-off was uneventful from Birmingham International Airport, and as they climbed through 17,000 feet over the Oxfordshire countryside, everything seemed routine. The cabin crew started to prepare a meal and drinks, and the captain loosened his shoulder harness and lap strap and settled down for the flight ahead. There was nothing to indicate that his life and the lives of his passengers were in imminent danger.

As the plane passed over Didcot, there was a sudden loud bang as the windscreen suddenly popped out and the surging decompression pulled Captain Lancaster out of the window, and the flight deck door off its hinges into the cockpit. Only the quick reflexes of a steward who rushed into the cockpit and grabbed the captain's legs, saved him from certain death. The aircraft rolled to the right as the disorientated co-pilot fought to gain control of the aircraft. Fortunately he was able to bring the plane into a rapid descent and make a safe emergency landing in Southampton.

Like more than two-thirds of aviation accidents it was not equipment failure or metal fatigue which caused the accident— it was human error. The maintenance engineer who had just replaced the windscreen had used the wrong bolts. Eighty-four of the bolts were each 0.026 of an inch less in diameter than required, and six others were of the correct diameter but 0.1 inch too short.

Was this a careless individual who was responsible for such an unforgivable error? No, it was one of British Airways' best mechanics, with many commendations to his credit. But he was placed in a situation where the human body too easily succumbs to inattentive failure. His rotating around-the-clock work schedule required him to replace the windscreen in the small hours of the morning, between 3:00 A.M. and 5:00 A.M., on his first night shift of the week, before his body could adapt to the night routine. As this book will show, under such circumstances, without special precautions, even the most responsible person can inadvertently fail.

THE IMPACT OF CHANGE

FOREWARNED is forearmed—as long as you don't bury your head in the sand. British readers, right now, have a unique opportunity to see what is in store for them as we race towards the Twenty-First Century, and a world where many people in critical jobs are required to work around the clock.

This book chronicles the next fundamental societal revolution—the conversion of the industrialized world into a nonstop Twenty-Four Hour Society. The enormous impact of this revolution has been sweeping across the United States of America at an accelerating pace over the past ten years. It is profoundly changing the nature of business and the lifestyles of millions of Americans. Because of the inescapable momentum of this revolution—propelled by technological innovation and economic imperatives—Great Britain and the rest of Europe cannot be far behind.

It is tempting to be complacent as the Twenty-Four Hour Society sweeps across the globe, and assume that just because you live in Godalming or the Lake District, what they choose to do in America or Japan, or even the factories of Birmingham in the middle of the night, doesn't affect you. But nothing could be further from the truth.

The people who can most impact your personal safety work long and irregular hours, often around the clock. It is not only the round-the-clock work schedule of the aircraft mechanic that places you at risk. The lorry driver looming up behind you on the motorway, the surgeon who is about to operate on you, or the electrician re-wiring the signals on your railway line are all liable to be fatigued and inattentive.

The human costs of this societal and economic transformation into a nonstop world—the fatigue, errors, accidents and ill-health—do not lie just in the future. They are already upon us. Already at least four million British citizens work outside regular daytime hours, about half of them on schedules which require overnight work. And the other 55 million are by no means free of risk.

In this book you will meet the fatigued British pilots who almost stalled a Lockheed 1011 Tristar while on final landing approach. You will learn how 35 people died at Clapham Junction because a signal repairman working excessive hours overlooked a simple safety measure. You will find out why a sleepy lorry driver cruising at 65 mph crashed into backed-up traffic on the M42 and killed six people without ever lifting his foot from the accelerator. And you will see the long arm of human error at Chernobyl, where human fatigue and impatience in the early hours of the morning resulted in a 10-fold increase in radiation over Cumbria and North Wales, making livestock unfit for human consumption.

But this book does much more than just alert you to the dangers. It shows how to control the risks and make the personal and business judgments needed if we are to benefit and profit from a nonstop world. European companies are starting to apply the new human alertness technology discussed in this book. They are learning how to profit from nonstop operations and reduce the risks. The nonstop global marketplace is now open for business. It will take, however, an enlightened British appreciation for the quality of life if we are to ensure that appropriate attention is paid to human characteristics and needs in a technology-driven world.

I

THE
SEEDS
OF
REVOLUTION

Chapter 1

Outpaced
by Our
Technology

EARLY ONE MORNING at three o'clock, as a DC-8 of a major airline was on final approach to Chicago's O'Hare Airport, the second officer (flight engineer) noticed that the plane was headed not for the runway but straight at the American Airlines terminal. He immediately alerted the first officer, who was flying the plane, but the first officer responded in a dazed, zombielike voice, "Everything is okay, I am on a vector to intercept the runway." Realizing that both the first officer and the captain were in a sleepy stupor, the second officer yelled out in desperation, "Where are you going?!" Fortunately, the captain then responded and landed the plane safely while the first officer sat beside him like a passive lump.

The first officer had sunk into an automatic behavior state in the netherland between sleep and wake. Like an automaton disconnected from reality or relevant data input, he could still follow through on the well-rehearsed automatic process of landing a jetliner. But incoherent and unresponsive, he was obviously a danger to all on board—a nightmare of human error propensity just barely averted.

On the other side of the world, the 290 passengers on another airplane were less fortunate. For below them, in the Persian Gulf, steamed the technological pride of the United States Navy, the USS *Vincennes,* a $1 billion Aegis cruiser—and no ordinary ship. The skipper did not scan the horizon from the bridge with heavy binoculars, but instead exercised command from the high-tech Combat Information Center (CIC), a windowless room

3

linked to the outside world through glowing computer and radar screens. Never before had a warship's captain had access to so much instant and accurate information or had so much technological power at his fingertips.

But all this technological prowess was only as good as the weakest link in the chain—the sailors who operated the technology. Unfortunately, the crew had been fatigued and stressed by frequent calls to general quarters battle stations. Each time the exhausted crew members went below decks to sleep, another small Iranian patrol boat was spotted carrying potential attackers, and, following combat procedures, again the crew members were called out of their bunks.

The path of the Iranian A300 airbus, as it climbed for its scheduled short trip over the Persian Gulf, was detected by the *Vincennes*'s Aegis system. Although the radar showed the commercial airliner climbing on a normal flight path, at least one fatigue-stressed CIC operator anxiously and repeatedly told the captain that the "target" was descending, as if to attack. Another key officer struggling against fatigue distorted the data to fit his preconceived perceptions of an attack scenario.

Despite accurate contraindications from sophisticated computers, the fateful decision was made to launch a missile, which intercepted the plane and sent 290 innocent people to their deaths. This disaster in turn spawned the retaliatory bombing of Pan Am flight 103 six months later, in which 270 people lost their lives as death and destruction rained down on the inhabitants of Lockerbie, Scotland.

NOT BUILT FOR A WORLD
WE HAVE DESIGNED

THESE ERRORS and accidents did not occur by chance. Myriad other, equally frightening mistakes happen every day. Each is the predictable outcome of failing to allow for human limitations in the nonstop world we have designed and of investing much more in the technology of machines than in the performance of people (Figure 1.1). The engineers, designers, and managers of our society have not yet learned this lesson. Indeed,

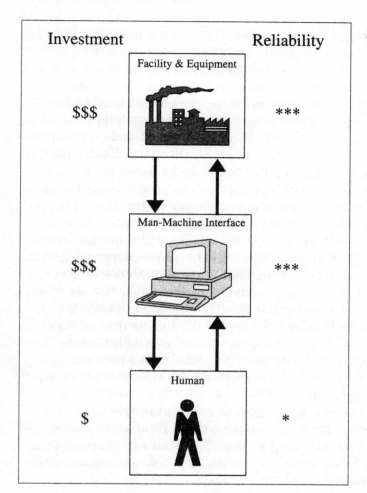

Figure 1.1

we continue to design equipment, systems, and procedures that can only increase the risk of catastrophic human error.

The crisis in human performance is hardly limited to the world of aircraft and radar screens. All across the nation in our finest teaching hospitals, the wonders of high-tech medicine are rendered useless by mind-numbed interns and residents working day and night on thirty-six-hour shifts. On our highways, long-haul truckers working fifteen-hour days are lulled into nonresponsive zombie states while driving huge vehicles at high speeds. And it is not a coincidence that the most notorious industrial accidents of our time—Three Mile Island, Bhopal, Chernobyl, and the *Exxon Valdez*—all occurred in the middle of the night, when those with hands-on responsibility were dangerously fatigued. The problem isn't limited to the control room, however. Management, too, fails to observe its own limitations, as, for example, in the *Challenger* disaster, when key NASA officials made the ill-fated go-ahead decision after working for twenty hours straight after only two to three hours of sleep the night before. Their judgment error cost the lives of seven astronauts and nearly killed the U.S. space program as well.

At the heart of the problem is a fundamental conflict between the demands of our man-made civilization and the very design of the human brain and body. Fashioned over millions of years, we pride ourselves as the pinnacle of biological evolution. But the elegant organization of cells and chemistry, structure and systems, sinews and skeleton, that is the human being, was molded in response to long-outdated design specs that we seem to have forgotten. Our bodies were designed to hunt by day, sleep at night, and never travel more than a few dozen miles from sunrise to sunset. Now we work and play at all hours, whisk off by jet to the far side of the globe, make life-or-death decisions, or place orders on foreign stock exchanges in the wee hours of the morning.

The pace of technological innovation is outstripping the ability of the human race to understand the consequences. Much like a toddler who is truly dangerous because his physical abilities have developed well ahead of his cognitive awareness, our society has reached a critical and hazardous stage of development. And the problem will get worse. Competitive pressures are forcing more and more businesses to provide their services twenty-four hours a day. Demands for increased returns on capital investment and lower costs are requiring most manufacturing plants to convert to around-the-clock, seven-day-a-week operations. International competition, made possible by the wizardry of modern telecommunication, is speeding up the pace of global trade and the demand for international travel. And the technology of fax machines and laptop computers seems designed to prevent us from ever escaping the fatigue of working at all hours of day and night—wherever in the world we may be.

Meanwhile, the trend is for ever more power to be given to an individual. Oil refineries are centralizing their computerized control systems so that a single operator is responsible for more distillation towers and cracker units. Chemical plants are converting from batch processing to continuous processing so that more chemicals can be synthesized with fewer employees. Trucking companies are adding double and triple rigs to trucks so that a single driver can transport more goods. Aircraft manufacturers are designing the next generation of superjumbos to transport more passengers but reduce the cockpit crew from three to two.

With more power comes more risk. Instead of only hundreds or a few thousand dollars being placed at risk from a single human operator error, now the costs are in the millions or even billions. A fatal heavy-truck accident typically incurs $1 million or more of liability; an error of inattention in a nuclear power plant, in a chemical plant, or on an oil tanker can create a $1 billion problem. The tab for Exxon from the *Exxon Valdez* oil spill, which the National Transportation Safety Board concluded was due to crew fatigue and inattention, has so far exceeded $3 billion, with more than $50 billion of claims in the legal pipeline.

Such risk gives today's chief executive nightmares. An error by a junior person earning one-hundredth of a CEO's salary and whom the CEO has never met can bankrupt the company and deprive the CEO of a job. Few companies have the resources of Exxon to pay out billions of dollars and survive, and even Exxon cannot do that often. Indeed, Richard Stegemeier, the CEO of Unocal, has recently circulated a videotape to his employees urging extreme care in safety-sensitive operations such as shipping or trucking oil, because it would take only one such major accident to bankrupt his company.

PERILS OF A NONSTOP WORLD

OUR TWENTY-FOUR-HOUR society has developed because of a set of technological and economic imperatives. The technology of telecommunications, satellites, and fax and computer networks means that we can pass a message or receive the answer to a question between Boston and Paris or London and

Tokyo faster than it once took to walk down the street to consult a neighbor in a medieval village community.

All that separates us now are the time zones—but therein lies the problem. Hustle and bustle, decisions, deals, and opportunities occur continuously, because it is always daytime somewhere in the global village. And those working in that daytime create a demand for instant attention from others who must work by night on the other side of the world.

This constant hive of activity is reinforced by the fact that the technology of our society, the machines and equipment, are designed to run continuously without concern for night and day. Indeed, the economics of production and capital investment are so much in favor of using assembly lines and processing plants continuously that large sections of the population now work at night or on rotating shifts. They in turn need services at all hours, and so yet another group of people is inexorably drawn into this world where nature's temporal order no longer rules.

All this might be fine if the human body were infinitely adaptable. But our patterns of sleep and wake, of digestion and metabolism, are governed by internal biological clocks, elegantly attuned to the patterns—of dawn and dusk, night and day—of a simpler era. They make us sleepy at night and alert during the day, and they regulate the activity of our digestive tract to be ready for the next predicted mealtime.

The twenty-four-hour society forces us to operate the human body outside the design specs crafted by prehistoric experience. The solutions we must seek require a rethinking of society, and a sophistication in biotechnology to adapt the human body to the technology and the technology to the human body. Indeed, we are in the midst of the next major societal revolution—a time of crisis and of opportunity.

SOCIETY IN REVOLUTION

SEVEN THOUSAND years ago, the agricultural revolution saw the hunter-gatherer in most civilizations become a planter-harvester. An incessantly mobile human society was thereby converted into one invested in hearth and homestead. Two hundred years ago the industrial revolution converted that well-refined agricultural society into one dominated by the dictates of industrial production. Now we face another such major revolution— the conversion of our world into a single technologically integrated, around-the-clock community.

Each such societal revolution was initiated by a set of unfulfilled needs and a set of innovations and inventions that responded to those needs. In the case of the agricultural revolution, population growth had strained the natural limits of food supply and made apparent the inefficiencies and uncertainties of the hunter-gatherer lifestyle. Responding to those needs was a stream of innovations such as planting, harvesting, breeding and raising livestock, home building, and defending territory. Similarly with the industrial revolution the inefficiencies and uncertain quality of widely distributed cottage industries could not meet the demand for goods. Innovations such as steam-powered engines, industrial machinery, fuel supplies and energy production, steel, city construction, mass-produced row housing, water supply, and sanitation were designed to solve the problems and needs of industrialization.

Our present revolution—the conversion of our world into a nonstop twenty-four-hour society—has developed in response to our need for ever more efficient production in a world with too-scarce capital resources to meet the needs of an explosively growing population. It has been made possible, like other societal revolutions, by a stream of inventions and innovations: Instant worldwide telecommunications that allow us to pick up a phone and speak to a person at any point on the earth's surface, whether on a vehicle, roaming the desert, or out at sea. Computers and microprocessors—a computer on every desk, every lap, and even every palm, tapping into vast resources of information and computing power. Automation of machinery that will run continuously for weeks or even years without human intervention or servicing. Jet and supersonic travel, moving people and freight twenty-four hours a day rapidly point to point across the earth's surface. A global economy and trading system, and the tumbling of major political boundaries such as the Iron Curtain. Surely the Bamboo Curtain must fall in not too many years; already sixty million Chinese in the provinces near Hong Kong are developing a market economy under the principle "the mountains are high and the emperor is far away." Modems and fax machines bridge the gap to the outside world more readily than the old lines of communication keep contact with the aging power structure in Beijing.

Societal revolutions have the habit of sneaking up on us. The agricultural revolution and the industrial revolution were each well under way before most people realized what was happening around them. To look back with the perspective of time and recognize how the world has changed is easy; to recognize the subtle seeds of change that will flower into a full-blown societal revolution is much harder. Only after many years, when the pundits and historians have put labels on a series of events and fundamental societal changes, is the societal transformation conceptualized as a revolution.

People who understand early that a fundamental societal revolution is beginning can make the right decisions, choose the right investments, and modify their lives and businesses appropriately. They place themselves in a position to benefit most from the positive aspects of the change and suffer least from the negative. But the vast majority of people remain deeply embedded in the paradigms and assumptions of the past era. They do not wake up in time to comprehend how their world has changed and do not recognize early enough how they should change their lives accordingly.

We are currently in the midst of a revolution as fundamental and as far-reaching as any previous societal revolution. This book chronicles this revolution and shows how to take advantage before the fact—where to invest, how to modify one's life and compete more effectively in one's business, and how to avoid the side effects that can rob one of health and financial security, and cause fatigue and stress.

This is not a simple "how to" book, although where to find specific necessary information will become clear. Rather, it is a road map, a big-picture look at the transition we are passing through, a glimpse of how the world will look in the rapidly approaching twenty-first century. We must understand the big picture to make the right decisions for ourselves, our families, our businesses, and our lifestyles.

Chapter *2*

Facing Up to the Challenge

I FIRST BECAME aware of the challenges and risks of working around the clock as a newly minted physician, starting my first job as a surgical intern. To my dismay I found that I was required to work shifts of up to thirty-six hours straight, spend hours on end in the operating room in the middle of the night, and make life-or-death decisions around the clock in a perpetual state of fatigue. After a few weeks, my colleagues and I were transformed from paragons of bright-eyed, enthusiastic energy into a state of exhausted, head-nodding stupor.

Despite the long hours and shortage of sleep, there were times when the energy came back and the fog of sleepiness rolled away—even when I had not slept for many hours. Somehow in the mornings and early evenings, energy and clarity returned for a while before the dense fog of fatigue descended once again.

The experience of the powerful cyclicity of alertness and sleepiness triggered my curiosity in the biological processes that determine when we are tired and when we are not. I decided after my internship not to continue my clinical residency but instead to sign up for the Ph.D. program at Harvard. I wanted to explore this trail of inquiry into what drives us to sleep at night and stay awake by day.

At that time, in the early 1970s, circadian (approximately twenty-four-hour) rhythms had been demonstrated in body temperature, hormone levels, and a wide variety of other bodily func-

11

tions, each with its own pattern of rising to a peak level at one time of day and falling to a low point at another. However, the biological clock that generated these rhythms had not been identified, and little attention was being paid to the implications of these circadian cycles for human health and well-being.

The next ten years were an exciting time. As I completed my Ph.D. and was invited to join the faculty at Harvard Medical School to build the Laboratory for Circadian Physiology, the scientific implications of circadian rhythms were rapidly becoming apparent. The biological clock that generates circadian rhythms was identified, and researchers around the world were making rapid progress in unraveling the secrets of biological timekeeping.

My colleagues Elliot Weitzman, Charles Czeisler (a young medical student at the time), and Richard Kronauer, and I set up the first laboratory in the United States to study the characteristics of the human clock. Working neck and neck with a research team in Germany at the Max Planck Institute, sometimes ahead, sometimes behind, we elucidated, step by step, the properties of the human biological clock, genetically enshrined more than one million years ago. We determined the mechanisms that govern the timing and length of sleep and wakefulness, and identified the location of the biological clock in the human brain. Most important, we learned how to manipulate the timing of this clock and how to reset the sleep-wake patterns of people who were out of synch with the world around them.

THE FIRST INDUSTRIAL GUINEA PIG

BY SOME QUIRK of fate, a story about our research on the timing of human sleep was picked up and published on the business page of the local Ogden, Utah, newspaper. The first thing we heard about it was a pink telephone-message slip that read, "I have a hundred shift workers who cannot sleep, can you help?"

The message was from Preston Richey. As production manager of the Great Salt Lake Minerals and Chemical Company in Ogden, he was responsible for the around-the-clock operations

that harvested the salts from solar evaporation ponds juxta-
posed to the Great Salt Lake. Day and night, huge front-end
loaders scooped the crystallized salt and dumped it into trucks
lined up in nonstop progression.

On further investigation we learned that Richey was battling
in his work force the problems of chronic fatigue, sleep deprivation, and sleepiness on the job. And no wonder: the employees
were on a weekly counterclockwise schedule that was equivalent
to spending a week in Utah, a week in Paris, and a week in
Tokyo, in an endless, jet-lag-inducing rotation.

Richey invited Chuck Czeisler and me to examine his operation. After arriving, we were briefed on how the operation ran
and the problems the plant was facing. As we sat around the conference table, the management team asked us if we could help.
We responded that almost anything would be better than the
current counterclockwise weekly rotation.

We drafted a game plan and presented it with great enthusiasm. Of course, the first question the plant manager asked, with
appropriate caution, was "Have you done this before?" Coming
from our academic ivory tower, we answered, "No, you are going
to be the first test site for our theories, isn't that great?" Jaws
dropped. As we subsequently learned, no manager wants to be
the first at doing anything—it is much safer to be the "nth." But
an academic research scientist is only interested in being the
first. There is no such thing as being the second person to invent
penicillin.

That the Ogden plant's management let us take a shot at fixing their work force's chronic fatigue problem is testimony to
their courage and our persuasiveness. We devised a brand new
biocompatible way of scheduling shifts around the clock and provided training programs for the shift crews. Inventing on our
feet as we went along, we devised ways to implement the
changes and solve the myriad problems along the way.

The results were dramatic. Productivity rose by an unheard-
of 22 percent as a tired work force was converted into a relatively energetic one. This productivity gain was seen in
truckloads of potash leaving the plant—the alert operators
became more efficient at operating the front-end loaders, fewer
loads missed the truck they were intended for, and the pace of

operations hummed forward at record speed. At the same time, employee turnover dropped markedly because the work force much preferred the new schedule. Health surveys showed that medical symptoms subsided and morale improved greatly.

The productivity gains added $800,000 to the plant's bottom line in the first year alone, without the purchase of any more capital equipment or the hiring of any additional employees. We were, of course, concerned initially that we might be seeing the so-called "Hawthorne effect," an increase in productivity caused just by the special attention given to the employees. But the increase was sustained year after year after we left the plant, when the special-treatment possibility no longer applied.

We published our results in the journal *Science* in the summer of 1982 and experienced our first splash of publicity. By that fall more than two hundred companies had called, wanting our "special medicine" for increased productivity, and we realized that we would have to set up consulting operations to address the need.

Over the subsequent years my consulting group, Circadian Technologies, has created strategies for many manufacturing plants, nuclear power stations, chemical plants, oil refineries, police forces, airlines, and a wide range of other special niches in the twenty-four-hour society. We have devised many different approaches to deal with the practical problems of the various types of around-the-clock operations. No longer must we deal with managers' concerns about being the first guinea-pig site; we can now assure them that they are indeed the "nth."

By wrestling with problems such as how to keep Federal Express pilots awake in the cockpit, train nuclear power plant operators to stay alert on duty, design duty schedules for the United States Navy submarine fleet, or reduce human fatigue on oil-well platforms or in General Motors' new Saturn plant, I have learned more about the realities of the twenty-four-hour society than I could have any other way. I have spent nights in the control rooms of nuclear power plants, oil refineries, and chemical plants; crossed the international date line in the cockpit of a commercial airliner; designed NASA space experiments; visited Pentagon command centers; and talked with innovators and leaders such as Fred Smith, the founder of Federal Express, and Neil Armstrong, the first man on the moon. There is nothing like hands-on industrial consulting to uncover the real issues in the various types of around-the-clock operations. I scrutinized the operations, warts and all, and I learned where the barriers and limits were by trying to set change in motion. Indeed, despite our many successes, we found that we uncovered more problems than we solved.

OVERLOOKING THE OBVIOUS

THE FIRST MOUNTAIN we had to climb was to convince managers that they had a problem. Managers of around-the-clock operations typically go home at the end of the day and often are not fully aware of how their facilities operate in the dark of night. They also do not personally experience the mind-numbing fatigue day in and day out that is generated by rotating hours of work and sleep. Even when managers show up at night, they don't see the true picture. Every plant I have visited has an informal warning system, usually triggered by security at the gate, and by the time the manager walks onto the floor, the lights are turned up and everyone is alert and attentive to the task at hand. Only by living at the plant night after night can one merge into the scenery and learn how the plant really operates.

If this is true for plant managers, think how far removed from reality are senior corporate management or the engineers who design the equipment that people use in around-the-clock operations. A few worked shifts around the clock early in their careers and have some idea of what such work is like, but most are unaware or ill informed and make judgment errors that would be unthinkable if they really knew what was going on.

Indeed, many costly industrial accidents occur precisely because intelligent and responsible people are overlooking the obvious. Engineers are building highly sophisticated man-machine gee-whiz interfaces for oil refineries and chemical plants and neglect to consider that they resemble a TV set in a dimly lit room. They expect the operator to remain alert and attentive even though it is 3:00 A.M., the program is boring—night after night it is the same rerun of numbers, charts, and data—and the operator may have been awake for nearly twenty-

four hours at a stretch. Just because the job description says
"thou shalt not fall asleep" doesn't mean that human nature of
the most obvious kind won't take over.

In talking with the senior management of around-the-clock
businesses and the engineers who design the man-machine interfaces in twenty-four-hour operations, I realized that pointing out
the obvious is not enough. Time and time again, I have been
offered any number of excuses or reasons for not investing in the
development of technologies that address the problem.

That blocks of the mind-set are a fundamental human frailty
was readily apparent from conversations with my wife, Dr.
Donna Moore-Ede, a skillful and effective clinical psychologist.
She sees as one of her greatest challenges the need to convince
her patients of the obvious. She learned many years ago that people simply cannot hear such truths until they are ready. They
will even say later, "Doctor, why didn't you tell me that," even
when they were explicitly told it time and time again. Because of
the inertia of their mind-set, they were just not ready to hear.

As I continued my rounds of twenty-four-hour businesses, I
did find champions who immediately grasped the issue and set
change in motion in their own companies. Over the past ten
years the momentum has grown, as the productivity, safety, and
health gains from addressing the problem of human frailty in a
nonstop world have become apparent. Now it is much more common for managers to be fully aware of the problems and to want
to cut straight to the possible solutions.

The last bastion of resistance seems to be the engineers who
design the man-machine interfaces and the equipment used by
around-the-clock workers. Their marvels of engineering and software represent years of investment, as the reliability and capabilities of their machines have been steadily perfected. Unfortunately, however, that technological commitment has resulted
in inertia in addressing the problem of human alertness.

THE BIRTH OF A NEW TECHNOLOGY

ANOTHER CHALLENGE we faced when consulting in the
early days was the enormous gap between research in the academic laboratory and the realities of the actual world of around-

the-clock operations. We constantly had to make huge extrapolations from basic scientific principles to tackle practical problems in the workplace.

To address this need for applied research, we spun off from Harvard Medical School to form a nonprofit research center, the Institute for Circadian Physiology. Sponsored by twenty corporate partners and by government and foundation grants, the mission of the institute is to develop the scientific basis for human alertness technology—to provide "Alertness When Required, Sleep When Desired."

The centerpiece of the institute is the Human Alertness Research Center. HARC contains a unique simulation of an around-the-clock workplace, complete with industrial control rooms, man-machine interfaces, and apartments where subjects can live and sleep when not on a work shift (Figure 2.1). This microcosm of an around-the-clock workplace and lifestyle has enabled us to re-create many of the problems of twenty-four-hour operations and to devise effective solutions.

Figure 2.1

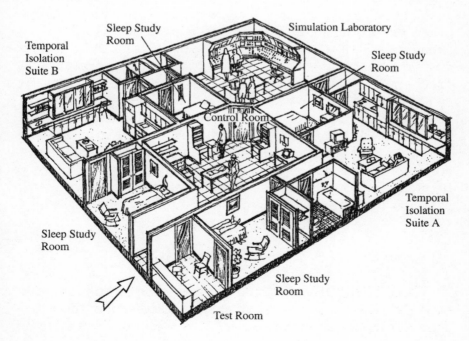

As a result of this research, the science of alertness, sleep, and biological clocks has matured to a new level of sophistication. Exciting new technologies are pouring out of the research labs, and powerful, safe techniques are becoming available to replace the ineffective and dangerous practices of the past. These new approaches will revolutionize the way we think about the timing of our lives.

The next three chapters introduce the science of alertness and the consequences of human fatigue. Then we will explore examples of around-the-clock workplaces, from the airplane cockpit to the nuclear power station to the hospital emergency room, where we all have a vested interest in how fatigue is managed. Next we will look at an innovative, emerging human alertness technology that will help the human body and brain cope with life in the twenty-first-century global village. Last, we will examine how each individual must adjust his or her lifestyle, and how each around-the-clock business must organize its operations, to cope with the realities of a nonstop world.

Chapter 3

Design Specs
of the
Human Machine

SOME YEARS AGO I had an opportunity to work with some of the engineers who were designing the crew cabin of the space shuttle. Having been involved in NASA biomedical experiments since the Apollo moon mission days, I was invited to serve on a NASA expert panel advising on human factors considerations for the shuttle. For a young scientist, to work with the best in American engineering—the men and women who had put human beings on the moon—was a thrilling opportunity.

As we looked at several possible crew cabin configurations, my biomedical colleagues and I strongly urged NASA to abandon one scheme, which we felt almost guaranteed that the crew's blood circulation would be drawn into the legs during landing, causing unconsciousness. Despite our protests, we were eventually told that the engineers had selected this design.

A year later I learned that the design had subsequently been abandoned in favor of one of the others we preferred. When I expressed delight that the biomedical concerns had finally been acknowledged, I was told the reason for abandoning the previously selected design: undue stress from G-forces would damage the wings. We got the design we wanted, but we had done nothing to alter the mind-set of the engineers!

In most cases of discrimination one group discriminates against another, but here we have a case of discrimination against ourselves. We treat our man-made machinery better than the bodies of the people who operate it. Machines are protected by operation manuals, warning labels, and training courses. Humans arrive in this world with no such protection. We tend to assume that people are adaptable, but that is true only within narrow limits.

A manager of an industrial plant, or a pilot of a plane or a NASA space shuttle, who operated a complex piece of machinery outside the specifications for which it was designed would be deemed reckless and irresponsible. Operating a machine outside its design specs, at too many rpm, too high a G-force, too high a temperature, or too low a pressure, creates unnecessary risks of breakdown and failure. Yet the most highly sophisticated pieces of machinery in that industrial plant, airplane, or space shuttle are not the complex electronic man-made systems, but rather the bodies and brains of the human operators or pilots.

People believe that machinery is inherently unreliable and must constantly be watched. Yet, with the advances in modern engineering, it is the human operator, more than the machine itself, that needs watching. We hire people to watch the equipment in our factories, nuclear power plants, and airplanes, but we don't have the equipment watch the person to ensure that he or she is awake and alert. We have the weak link of the chain watching the strong link, but not the strong watching the weak—the fallible watching the infallible, rather than the infallible watching the fallible.

Because we have been *machine-centered* in our thinking—focused on the optimization of technology and equipment—rather than *human-centered*—focused on the optimization of human alertness and performance (Figure 3.1), the reliability of machines has grown enormously in the twentieth century, while human reliability has tended to decline (Figure 3.2). Human error has become the problem of our age because the trade-offs and compromises, made to ensure the technological achievements of the modern world, have not taken into account the design specs of the human body. Creating and installing a human-centered technology to redress the balance will be one of the most important challenges of the twenty-first century.

The shocking truth is that we know far less about the design specs of the human being than we know about the hardware and software he or she operates. And we tend to abuse those design

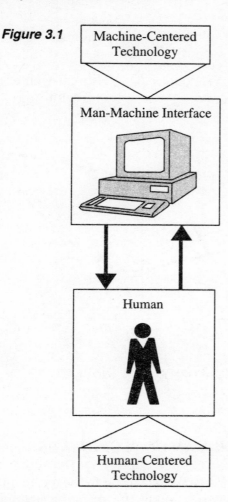

Figure 3.1

specs, particularly in around-the-clock operations, which are the core of our twenty-four-hour society. People don't release smoke, grind gears, or have pieces fall off; but their equivalents—fatigue, error, injury, and ill health—do result in failure and breakdown.

Let us look carefully at this human machine, particularly at the features that cause us trouble in our nonstop world. For we are beautifully engineered to address the design specs of a more primitive world, a world governed by the dictates of day and night on a spinning planet, rather than the continuous demands of efficient technology.

Figure 3.2

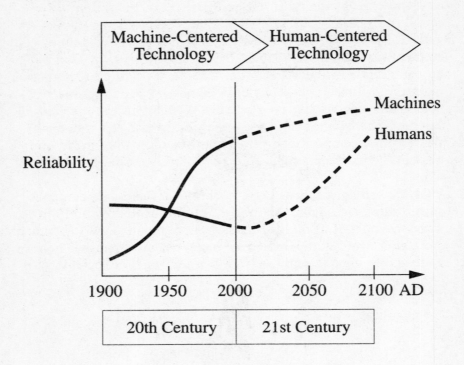

CYCLES OF HUMAN CONSCIOUSNESS

THE HUMAN HABIT of sleeping for seven to eight hours or more at a stretch is unusual in the animal kingdom. Consolidating all sleep into a single block and maintaining sustained wakefulness during the day is a particularly human trait. Most animals catnap, take their sleep and activity in smaller doses, and sustain any particular behavior for shorter spans of time. Some vestiges of that behavior still lurk in the cogwheels of humanity's sleep-wake apparatus—a trait we will return to later because we can exploit it to our advantage in our newly nonstop world.

Why does the human brain consolidate all sleep and all wakefulness into single blocks? This is the fireside or barroom discussion of the scientist-philosopher. Might sustained alertness have enabled the human being to rise above all other species, think great thoughts with sufficient concentration, solve complex problems with sufficient time, complete great tasks with this gift of focused attention? And might consolidated sleep bouts of eight hours have emerged as an effective survival behavior that kept our ancestors out of trouble at night by suppressing human curiosity and the urge to explore at dangerous hours? Our species has poor nocturnal vision and poor senses of smell and hearing as compared to animals that prowl by night. We were not competitive in the primitive nighttime world and were liable to become fodder for saber-toothed tigers and the like.

Myth, religious taboos, and teachings of the elders can pass on the same messages, warning of the threat to life and limb of wandering about at night. But nothing dampens a curious spirit more effectively than a biological urge to sleep. What we normally experience at night is a shutting down of biological sys-

Figure 3.3

tems, a fundamental state change of the brain and body. The
temperature of the body core drops to lower levels (Figure 3.4),
hormonal patterns change, and neuroendocrine and neural messages readjust the behavior of millions of metabolizing cells. We
sink into slumber, into a pseudohibernation, while the dangerous nighttime hours tick away.

Then in the predawn still of night, an hour or more before the
first rays of the sun creep over the horizon, long before we
finally awake, hormone levels start surging in the bloodstream
and the temperature of the body core starts rising. Heat-producing metabolism is switched on, and skin heat loss is shut down
by diverting the bloodstream away from the skin's surface. By
the time the human brain awakes, body temperature and other
essential functions are reaching daytime levels of activity. We

Figure 3.4
The Circadian Rhythm of Body Core Temperature

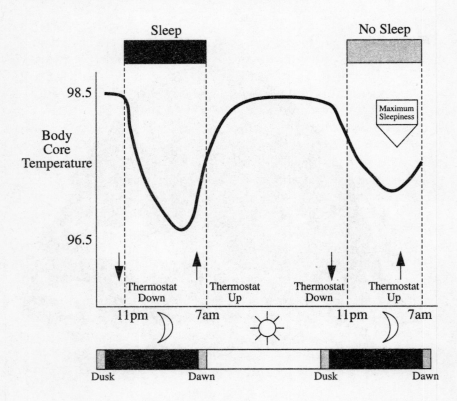

are ready for the day ahead, programmed by internal automatic systems for success in a dangerous primitive world.

We are sustained during the day by brain mechanisms that suppress sleep and keep our bodies operating at full throttle. Then again as the next night approaches the systems are readjusted for slumber in the night ahead.

Sleep itself is not a simple on or off state. We cycle through quite distinct stages, each characterized by changes in the electrical activity of the brain (Figure 3.5). This can be measured by attaching small wires to the scalp that pick up the rhythms in electrical potential coursing across the brain and spill them out onto a recorder called an electroencephalogram (EEG).

When we are alert, our electrical brain waves measured by the EEG are fast (13 to 35 cycles per second) and random, but as we become drowsy the brain waves start to slow into a regular "alpha" pattern (of 8 to 12 cycles per second), which is exaggerated when we close our eyes. Then we start slipping into a semiconscious state, called stage 1 sleep, characterized by a further slowing of the brain-wave rhythm to the "theta" range of 3 to 7 cycles per second, but in which we remain vaguely aware of our surroundings. We may even convince ourselves that we are still awake and in control of our consciousness—a dangerous misperception if we are on duty in a critical job. From stage 1 we progress unknowingly into stage 2, a light level of sleep in which bursts of electrical activity called K-complexes and sleep spindles intrude into the EEG. Finally, after 30 to 40 minutes, we sink into the deepest stages of sleep (called stages 3 and 4) in which the brain waves slow down to 0.5 to 2 cycles per second and are magnified in their amplitude. These deep slow waves are called delta sleep.

We do not linger for long in delta sleep. In fact, we tend to oscillate in a 90- to 100-minute cycle between the lighter and deeper stages of sleep (Figure 3.6), interspersed with bouts of dreaming in which the brain waves suddenly speed up to an awakelike pattern and the eyes start moving rapidly from side to side. This stage, called rapid eye movement (REM) sleep, may occur four or five times a night, building in duration as dawn approaches. Thus one may have four or five distinct dreams per night.

Figure 3.5
Electroencephalographic (EEG) characteristics of the human sleep stages. Characteristic changes in EEG waveforms are seen as a person moves between wakefulness and the various sleep stages.

Awake—low voltage—random, fast

50 µV

1 sec

Drowsy—8 to 12 cps—alpha waves

Stage 1—3 to 7 cps—theta waves

Theta Waves

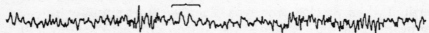

Stage 2—12 to 14 cps—sleep spindles and K complexes

Sleep Spindle

K Complex —

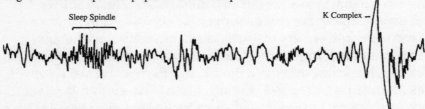

Delta Sleep—1/2 to 2 cps—delta waves >75 µV

REM Sleep—low voltage—random, fast with sawtooth waves

Sawtooth   Sawtooth
Waves   Waves

Figure 3.6
The sequence of sleep stages throughout the course of a normal night's sleep. The black bars represent the occurrence of REM sleep.

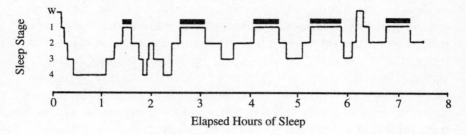

The stage of sleep from which you awake determines your condition on arousal. If you are in the deepest stages of delta sleep, you will feel groggy and disoriented on awakening, and suffer significant "sleep inertia" or impaired functioning for ten or twenty minutes or more. If you are in stage 1 or 2 sleep, you are much more likely to wake up alert and refreshed. Likewise you are more likely to remember a dream if you awake during a REM sleep stage. This information will prove useful later in this book as we discuss solutions for the modern world which optimize our ability to perform at our best, by ensuring that we awake from the optimal stage of sleep.

Our daily pattern of sleep and wake, metabolism and deactivation, that evolved in response to the globe spinning on its axis once per day is not inflexible. It was engineered to take care of such challenges as the gradual waxing and waning of day length and night length with the seasons. Although at the equator these do not change, in temperate climes the variations can be striking. Summer days may be many hours longer than those in winter. In the absence of such confusions as electric light, the timing of sleep and wake automatically adjusts to the season's changing hour of dawn.

Also built into the intrinsic pattern is a siesta time dip in early afternoon, when the propensity to sleep is not so deeply suppressed as at other times of day. Some people and some cultures take advantage of this built-in program, enabling them to reduce the length of nighttime slumber. Perhaps the most famous example was the British prime minister Winston Churchill, who during World War II always seemed to be awake whether it was just after midnight or 4:00 A.M. His secret was a two-hour nap in midafternoon.

This is fine if a person's culture, or status in society, enables napping in the afternoon. But most people have to toil away during these hours, often with less effectiveness and sometimes at more risk of danger. The postlunch dip, which is inaccurately named because it occurs whether or not one eats lunch, places one in danger because alertness is compromised and sleepiness intervenes. The postlunch hours are very dangerous on the highway, adding a strong blip to the accident statistics.

Still finer tuning of the system of sleep and wake is the pattern of the basic rest activity cycle (BRAC), a 90- to-100 minute oscillation throughout day and night. Just as sleep stages change throughout the night, so does the level of alertness vary during waking hours. The state often described as "bright-eyed and bushy-tailed" comes in bouts, as do surges in creative thought and concentration. In between are times when the circuits complete more slowly, when the vestiges of a catnapping ancestry show through. This BRAC pattern is most pronounced at night, when you would normally be in bed asleep. People sitting at the process control terminals of industry suffer strong intermittent bouts of intense sleepiness alternating with times when the fog clears and thoughts of sleep recede.

TRADE-OFFS AND COMPROMISES IN BRAIN PERFORMANCE

A LOOK AT the compromises made in the design of the human machine makes one realize how great the survival pressures must have been. During nighttime hours, and to a lesser extent during siesta hour, most types of human performance, whether manual dexterity, mental arithmetic, reaction time, or cognitive reasoning, are significantly impaired. Each type of performance typically heads downward at bedtime and then recovers during that internal reveille at dawn (Figure 3.7).

We are not well equipped to be aroused, even by danger, in the middle of the night and have our wits about us. This observa-

Figure 3.7
Circadian Rhythms in Human Performance

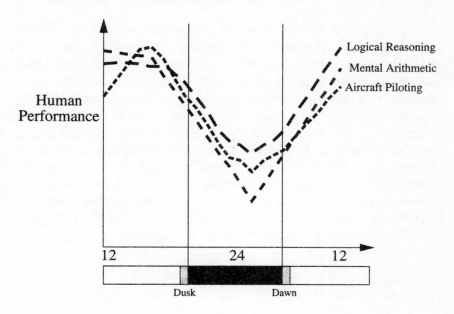

tion has not been lost to masters of the art of warfare, or to police making a predawn raid on a criminal at home. The grogginess, disorientation, and slow reflexes make the victim no match for the attackers with their adrenaline already pumped by anticipation.

Apparently the value of the nighttime block of sleep was so great that the evolutionary benefits outweighed the risks of nighttime deactivation. Fewer people lost their lives from being attacked while asleep than by wandering about at night. The recovery of brain functions achieved by that block of rest was more essential than the risks of switching off the computer.

This is not to say that we cannot function at night. Of course we can. But our physiological design compromises the brain's performance in nighttime hours. We are more prone to human error, to making that dumb mistake that can be so costly in today's nonstop world.

MEANWHILE, DEEP WITHIN THE BRAIN

THIS PRECISE TIMING of body functions, this ebb and flow of conscious and metabolic states, is not a chance occurrence. Instead it is the product of precise engineering, fine-tuned through millions of years of evolution to meet the demanding design specs of a very competitive, but predictably periodic, primitive world.

Only in recent years have we come to learn how elegantly nature engineered the clock that controls the systematic timing of sleep and wake. One tiny sliver of brain tissue, less than a millimeter across, less than the size of a pinhead, regulates the timing of our bodies. Within this sliver lies a biological clock that keeps track of the time of day, and seasons of the year, and marches our bodies and brains in step (Figure 3.8).

Figure 3.8
The biological clock regulating daily cycles of alertness and sleep is the suprachiasmatic nucleus (SCN), a small (⅓ mm) cluster of nerve cells in the hypothalamus, one of the oldest parts of the brain. A special nerve pathway (Retinohypothalamic Tract) conveys information about light and darkness in the environment from the eyes to the SCN.

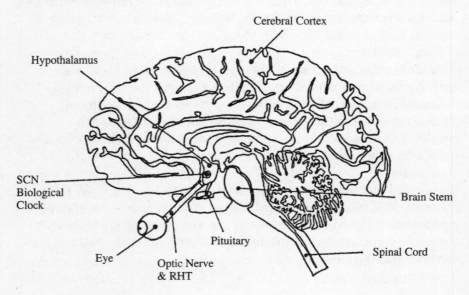

The small cluster of nerve cells that forms the biological clock is called the suprachiasmatic nucleus (SCN). Its name derives from the location of the SCN, just above where the broad optic nerve trunks cross over each other in a massive railway switching yard (the "optic chiasm") on their way back from the eyes to the visual centers of the brain. The SCN clock also receives information about light and dark from the eyes, but it has its own dedicated pathway of nerves, the retino-hypothalamic tract (RHT), that is separate from the main nerve bundles carrying visual information to the brain.

We know that the SCN is a biological clock, because when it is destroyed in an experimental animal by surgical pinpoint lesions of the brain, rhythms in sleep and wake, and many other rhythms, fade away. Interestingly, the animal, minus its SCN, runs, eats, and drinks the same total amount each twenty-four hours, but these activities are now randomly distributed throughout day and night. Similarly, patients with small tumors in that area of the brain become catnappers throughout day and night, never able to sustain prolonged bouts of alertness or sleep.

The SCN contains within it all the clockwork needed to sustain a circadian rhythm. This can be shown by removing from an animal a small slice of brain tissue containing the SCN and placing it in a dish containing nourishing fluids. The isolated SCN will continue to generate a regular circadian rhythm all by itself. The clock just keeps on ticking.

To show that the SCN, despite its small size, is the master pacemaker orchestrating the body's circadian rhythms, scientists have experimentally transplanted the SCN from a donor animal into another animal whose SCN has been destroyed. Such experiments have demonstrated that rhythmicity is regained by the recipient animal but with a schedule now determined by the time on the donor's clock.

It was not always obvious that circadian rhythms would prove to be generated by an autonomous clock within our brains. Before the SCN was found, it was reasonable to believe that this timekeeping function might be located in many of the body's cells, or even that time might be conferred passively from the environment so that, for example, darkness triggered sleep and light triggered wakefulness. Why did nature devise such an elaborate solution? The answer, as might be expected, lies in the advantages conferred by having your own clock.

PREDICTING THE FUTURE

SUCCESS IN LIFE goes to those who best predict the future, whether they saw the potential of IBM or Xerox stock before each became a household word or they bought a house before the last real estate boom.

The same is true in nature. Much is unpredictable, but some events you can rely on, such as the rising and setting of the sun, the ocean tides, and the seasons of the year.

Having a biological timepiece confers a decided advantage, enabling the animal to prepare in advance for a predictable event such as dawn or dusk. Each may be associated with the appearance of food (the early bird catches the worm) or of predators—depending which side of the biological equation you are on.

Advance warning provided by a clock is as important as having a spatial three-dimensional memory. To find something dynamic in nature, such as a juicy worm, that early bird needs to know not only *where* the worms will come out of the ground but also *when*; otherwise it would waste an inordinate amount of time hanging around waiting for worms.

A biological clock makes possible great efficiencies of time and energy. It allows risks to be reduced in an uncertain world. For when one is out there hunting, one is also at risk of being attacked by one's own predators. Forays into a primitive natural world needed to be swift and directed, precisely timed for maximum advantage and minimum risk. Journeys home to safety had to begin in adequate time before dusk; outbound trips to feeding sites needed sufficient time for the journey.

Why not do all this by the positioning of the sun in the sky? Anyone who has lived in England or in the Pacific Northwest knows the answer to that question. When clouds are socked in, the sun's position in the sky is an abstract concept, a wishful thought.

In recent years we have learned more about the essential role the clock plays in internal metabolism. It turns out that switching the body on in the morning is a slow process. Warming up the temperature of the body's thermal mass takes an hour or more; synthesizing the proteins to make the transporting pumps in cells can take a couple of hours. Our biological clock ensures that the internal reveille can start sufficiently in advance, before dawn, before arousal, so that we are ready to go when we wake.

It may seem hard enough to get going in the morning as it is. However, when you wake up two hours earlier than normal, to catch a plane or to make a special trip, your groggy, disoriented feelings are symptoms of the biological clock not having had time to prepare your body for the day ahead.

BALANCING REACTION AND PREDICTION

THE CELLS OF the human body are bathed by supporting fluids that are carefully maintained within a narrow range of acidity, temperature, salt concentrations, and oxygenation. Elaborate regulatory mechanisms exist to maintain careful control over levels of calcium, potassium, sodium, and so on. Layer upon layer of protective devices have evolved to buffer us from a highly variable world. This internal constancy, called homeostasis, is maintained by reactive systems. Body temperature too high? Switch on blood flow to the skin surface to flush the skin and lose heat and activate the sweat glands. Body temperature too low? Switch on heat-producing metabolism, start shivering to raise muscle temperature, shunt the blood away from the skin.

These systems continuously sense body temperature and thousands of other chemical and physical variables, and they leap into action at any time of day or night to bring these essential functions back within preset limits. They work quietly and automatically without conscious thought or decision making by the individual they protect. Indeed, as has been pointed out by the eminent physiologist Walter B. Cannon, "nature saw fit to place these essential functions beyond the caprice of an ignorant will." So well did Cannon convey his message of internal constancy that the medical profession has clung to the notion that body functions do not vary when one is at rest and in good health, and that deviations occur only in disease or when the system is challenged. Fortunately, these assumptions are changing with the growing appreciation of biological clocks and how they govern hormone levels and the effectiveness of medications and treatments with time of day. But the pace of acceptance is slow, and valuable new cancer treatments based on time of day go underused because the traditional scientific paradigm of an internal steady state remains strong.

The body strikes a compromise between the reactive homeostatic systems and the predictive modulations of physiological function provided by the biological clock. Thus body temperature is tightly regulated by reactive homeostatic systems throughout day and night, but the biological clock adjusts the set point, the ideal temperature that the reactive systems seek to achieve. Thus we can leap out of bed and respond to danger in the middle of the night, or stay awake all night at the controls of an airplane. We can eat a four-course dinner at 3:00 A.M. body time when we arrive in a different time zone. But we can achieve none of these functions with the same efficiency—or lack of dyspepsia—as we can when in synch with the dictates of our biological clock.

WITHOUT KNOWLEDGE OF TIME

THE FIRST STUDIES of the properties of the human biological clock were conducted in underground caves in France and in abandoned military bunkers in Germany. Volunteers lived for weeks or months without knowledge of the time of day. Elaborate ruses were constructed so that temporal information was not conveyed, newspapers were old, food deliveries were random, and even the research technicians were conscripted on an irregular routine.

The results: Without knowledge of day and night, people still rose out of bed; had breakfast, lunch, and dinner; and then retired for the "night." However, their daily rhythm systematically lengthened usually to approximately twenty-five hours. Their sleep and wake drifted across the day, lagging an hour per cycle, until after a couple of weeks their day was our night (Figure 3.9). The clock kept on ticking, of course, and in another ten days or so they were back in synch with the rest of us.

Such studies provided the first evidence for the intrinsic natural periodicity of the human clock. Far from a passive response to day and night in the environment, this clock had a life of its own. For example, an inanimate object such as a brick in the sun might have a daily rhythm in temperature because it warms up as the day progresses and then cools down at night; but if the brick is placed in a room at constant temperature, no self-sustained rhythm of temperature would persist. In contrast, a human body, with its own clock and temperature regulation, would heat up and cool off on its own cycle, even in that constant-temperature room. Furthermore, in the absence of environmental day-night cues, the body-temperature rhythm cycles

Figure 3.9

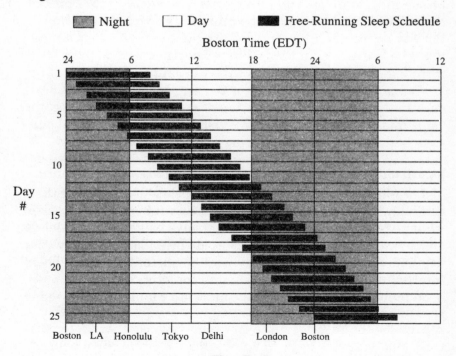

Time Zone

with its own natural twenty-five-hour rhythm for which there is no counterpart in nature.*

Under normal conditions, of course, the intrinsic twenty-five-hour cycle of the human biological clock is automatically, without our knowing it, reset forward by an hour each morning by the light of dawn so that it is restrained to the twenty-four-hour schedule of day and night. This process of synchronization is called "entrainment," and the time signals in the environment that do the entraining are called "zeitgebers," a German term meaning "time-giver." For most animal species, including humans, the most important zeitgeber is the environmental light-dark cycle. We will return to this subject because it offers great promise for control of the clock.

When we set up the first human time isolation laboratory in the United States in the mid-1970s, we did not need to resort to caves or bunkers. By then it was known that elaborate geophysical signals would not entrain the clock, although there seems to be some slight modulation from electromagnetic cues. Perched atop a hospital, we were in the middle of the hustle and bustle of the city twenty-four hours a day. Ambulance and police sirens could be heard within the isolation apartments, but these intruded equally whether it was day or night.

We had to take our own steps to exclude the environmental cues that might give our experimental deception away. Technicians and researchers who were visiting the experimental subjects had to refrain from wearing a wristwatch or offering salutations such as good morning or goodnight; males always had to be freshly shaven or permanently bearded.

The effectiveness of our deception became clear during the interviews at the end of each study. Most subjects kept track of what they thought were the passing days by keeping a calendar on the wall. When asked, however, at the end of the study what day and what time it was, they were consistently wrong. Their internal clock had slowed down, and sometimes they had even been lapped by environmental time. One young man refused to

---

*Yes, the lunar tidal rhythm has a 24.8-hour cycle, but a sufficient number of human "free-running" studies have been done to prove statistically that the human rhythm is not linked to the 24.8-hour lunar day.

believe that he had lost an entire day. We could not convince
him. Even showing him that day's *New York Times* was no use—
he was convinced that we were masters of subterfuge and had
printed a special copy with a future date. Finally, we had him
call his mother, who confirmed that indeed we were correct!

From these studies, the picture emerged of a robust system
driven by more than one clock. The fundamental metabolism of
the body, marked by the core temperature and the waxing and
waning of the hormone cortisol, marched with a precisely regulated rhythm, usually between twenty-four and twenty-five
hours in cycle length. The sleep-wake cycle was more elastic,
slipping occasionally into thirty-hour, forty-hour, and even fifty-
or-more-hour days. Thus subjects could accumulate weeks of lost
days over the course of one of our experiments.

Along the way we discovered a new method to lose weight,
unfortunately far too impractical for normal use! The subjects so
completely believed they were living a regular day that, when
they unknowingly drifted into a fifty-hour day, they continued to
eat three regular meals—of normal size, but spread, of course,
over twice the length of a normal day. They could not figure out
why they were shedding the pounds.

During these long days, sleep as well as wakefulness spread
out, but the normal 30 percent of time asleep was preserved.
Sleep now stretched out to fifteen, twenty, or even more hours. It
caused some consternation in the lab the first time this happened—had the subject sunk into coma? But he awoke
refreshed, believing that he had had a normal night of sleep.

One of the early lessons was that the length of sleep was very
dependent on the precise biological-clock hour a subject went to
bed. Sleep had regular behavior, especially when extraneous distractions were removed. Going to bed for several days at a regular time gave the person a certain baseline length of sleep,
which of course varied somewhat between individuals. Going to
bed later was not compensated equally by extended sleep. The
biological clock chimed in and woke the person near his or her
regular wake-up time, so the total sleep length was reduced. The
later and later the bedtime, the shorter and shorter the length
of sleep achieved—up to a break point in the morning of the next
day. But past that point, suddenly the person switched to very

long sleeps all the way through to the next morning, with the length of sleep again falling systematically in experiments in which the bedtime approached the regular bedtime the subsequent day. However fatigued the subject, the key determinant of sleep was the time on the biological clock.

CONFLICTS BETWEEN BIOLOGICAL AND ENVIRONMENTAL TIME

THE ROBUST intrinsic beat of the human clock is what causes problems in the work and travel schedules of the twenty-four-hour society. When environmental time cues are suddenly shifted by work schedule or transmeridian jet travel, the sleep-wake cycle and other rhythms of the body adjust only slowly. It may take several days, or even a week or more, for full adjustment to occur. In the meantime sleep, alertness, digestion, and metabolism are out of step, and fatigue, human error, and ill health may result.

Having an intrinsic twenty-five-hour period means that we are programmed with a steady westward drift in the absence of time cues. This is what makes westbound jet travel across time zones usually somewhat easier than eastbound (Figure 3.10). If on Monday the biological clock is on Boston time, for example, on Tuesday it is on Chicago time, Wednesday on Denver time, and Thursday on San Francisco time, and by the weekend you are on Honolulu time, without your clock having to adjust at all. Of course, exposure to zeitgeber time cues at one's destination makes the clock adjust faster than this, but the intrinsic natural drift helps the speed of transition whenever you are traveling west.

The natural drift westward in the absence of day-night cues also influences the range of day lengths to which the human body will entrain. Although it depends on the strength of the time cues, the typical range is plus or minus two hours on either side of the intrinsic near-twenty-five-hour rhythm. Thus a person can keep in step with down to a twenty-three-hour artificial day, or up to a twenty-seven-hour day.

The challenge to our limited range of entrainment comes not from natural variations in day length with the seasons but

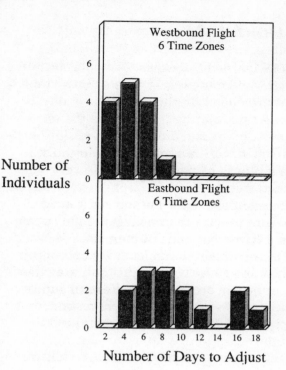

Figure 3.10
Adjustment Rates of Different Individuals' Biological Clocks to New Time Zone

instead from man-made schedules. Take the submariners of the U.S. nuclear navy as an example. They live for months at a time below the polar ice or wherever else they choose to lurk. For reasons of impenetrable logic based on human factors research that says a person cannot be effective manning a sonar for more than six hours at a time, and because the subs have room for only three crews, the crew (but not the officers) has characteristically been placed on an eighteen-hour artificial day-night cycle (after all, $3 \times 6 = 18$). But an eighteen-hour day means that the sailor must go to bed six hours earlier every day and wake up six hours earlier. This is equivalent to flying across the six time zones between New York and Paris every single day—an adjustment far greater than the human biological clock can achieve. We have not yet been able to convince the navy to change this pattern, with its unnecessary induction of fatigue—indeed, we found that the traditions of the submarine navy run very deep!

LET THERE BE LIGHT

ANOTHER LESSON learned from these temporal isolation experiments was that light and dark were the most important environmental cues that provided knowledge of time of day. It had been known for a long time that light cues were the most important zeitgeber in most plant and animal species, but previous studies had not confirmed this in humanity. Eventually, it was discovered that the strength of the light cue was very important in humans. In the early German bunker experiments, reading lamps were allowed at any time the subject wished. When we were much stricter about the provision of light in our studies—lights out meant lights out—the human clock showed evidence of conforming to the imposed artificial day and night.

That humans are much less sensitive to light than are other animal species has subsequently become clear. Regular indoor light intensities have only relatively weak influences on our clocks, so we are mostly protected from all but the influence of bright natural light. I will return to this later, for placing "sunlight" in a box has become a way to bottle the genie, to control the clock.

We now know that our bodies evolved special systems to precisely synchronize the biological clock with the timing of dawn and dusk. Special cells in the retina at the back of the eye detect the brightness of light falling on the eyes and route the information back to the SCN's biological clock. This is done by increasing or decreasing the bursts of firing up the RHT bundle of nerves from the retina to the SCN.

The clock behaves very differently in response to light signals depending on the time of day or night. During most of the day, the SCN clock virtually ignores light signals, but come evening, light signals (if sufficiently bright) act to delay the timing of the clock (sending it westbound). The distance shifted depends on the timing and the intensity of the light. Progressively, up to about 4:00 A.M. (presuming a normal sleep-wake schedule), the shifts get bigger and bigger the later the light is delivered. Then past the critical transition point in the middle of the night the shifts suddenly change from big delays to big advances (eastbound shifts). Progressively later still the advance shifts are reduced in size until, an hour or two past dawn, the clock is no longer responsive to light (Figure 3.11).

This so-called phase-response curve to light is fundamental to understanding how biological clocks synchronize to their environment. Fly westward and the delayed sunset means that light falls on a phase of the SCN clock that shifts it westbound. Fly eastward and early morning light from the earlier dawn advances the biological clock in synch. Sounds easy, doesn't it? Unfortunately, it sometimes isn't so simple in practice.

For example, for some reason airlines almost always schedule their flights from New England eastbound to Europe to take place overnight. If flying eastbound from Los Angeles to Boston, a plane covers a not dissimilar distance and takes barely an hour or two less. Yet on that trip all but the most penniless or most harried red-eye travelers fly during the day. In contrast, when flying eastbound to Europe, travelers leave early in the evening and often arrive at 5:00 or 6:00 A.M., which is midnight to 1:00 A.M. East Coast U.S.A. time.

Therein lies the problem. Walk outside in the bright sunlight

Figure 3.11

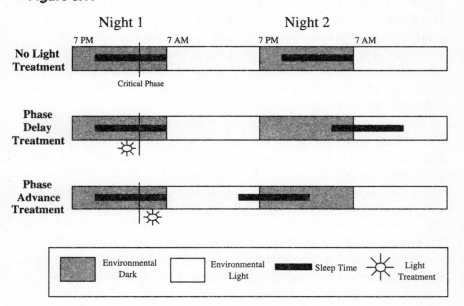

when you get off the plane in, say, Madrid, and your biological clock gets a healthy dose of light—but it is before 4:00 A.M. East Coast body clock time. The light signal therefore hits your biological clock at the phase when westward, not eastward, resetting occurs. This jolt of light will therefore send your body clock merrily on its way toward Hawaii time, just when you need to adjust to the schedule in Madrid. This is scarcely a help to your jet lag! If, however, you wait until 9:00 or 10:00 A.M. local time before exposure to bright outdoor light, the light will boost your biological clock eastward and help you with your adjustment to life in the new time zone. Such strategies form an important part of human alertness technology and will be covered more fully later in this book.

RESPECT FOR THE INDIVIDUAL

MORE THAN five hundred people, men and women, young and old, have now been studied in temporal isolation in the labs around the world. Earlier concerns that we were just seeing strange phenomena in people strange enough to agree to live in a cave or bunker for a month have been laid to rest. The rules of the human clock have been repeatedly confirmed. There are, of course, differences between individuals. Some people suffer greatly from jet lag; others don't have many symptoms. Part of the difference is the strategies people use to deal with it, but very real physiological differences between people account for most of the varying reactions.

Some people function very well on five hours of sleep a night, and others need up to nine hours. Some people are early-morning types ("larks"), who leap out of bed in the mornings, do their best work before noon, and become socially moribund in the evening unless they are reinforced by a nap. At the other extreme are the evening types ("owls"), who cannot get their motor running in the morning but are going great guns past midnight. Most people lie somewhere in between the two extremes.

Throughout history, people have interpreted one set of sleeping patterns as slothful, another as indicative of industriousness. In the words of Dr. Samuel Johnson, as recounted by

Boswell: "I have all my life long, been lying in bed until noon; yet I tell all young men, and tell them with great sincerity, that nobody who does not rise early will ever do any good."

No set of solutions for our twenty-four-hour society will be effective or acceptable if they do not address these important individual differences. For each of us to know our clock type if we are to work in the twenty-four-hour society will be just as important as it is for us to know our blood type if we are planning to have a major surgical operation. And it is equally vital for our society, if we are to avail ourselves of the best talent, to make sure that we build a world compatible with all human clock types.

Chapter 4

Alertness:
The Achilles Heel of a Nonstop World

WHILE VISITING a major chemical plant run by one of my clients, I was taken to admire the ultramodern control room. Intricate color graphic displays on the monitors were watched by highly trained technicians as the plant ran through its continuous processes. This was impressive high tech at its best. It would, and did, make a wonderful color photograph for the company's annual report, to assure stockholders that this company's facilities were state of the art.

Involved in some training with the night crew, we hung around and chatted with them, about baseball, the weather, their jobs. As night fell and the managers and engineers left, the room lights were switched off "to rest the eyes"; we sat there in our comfortable chairs, the room lit only by the dim glow of multicolor cathode rays. The room temperature was also adjusted up a notch as the crew settled down for the night. It was peaceful and quiet, save for the soporific hum of the computers. The plant was smoothly operating, splitting molecules and purifying them, filtering and storing them away.

It was so cozy in there that we wondered how they kept alert throughout a twelve-hour night shift. Curious, we asked what it was like to work with this equipment all night long. "Easy," the crew members said, "compared to the old days." "I just set this baby up, pull my cap down over my eyes and take in some zzz's. It wakes me if it needs me." They weren't even bothering to prop their eyes open with coffee for the long night ahead.

The next day, as we discussed with the management team

44

and systems engineers a new control room they were installing, we asked what the job task requirements were for the control room operators. "The job is to continuously and intently monitor the information on the screens all night long," they replied. "It is much more efficient and safe if they closely manage the process rather than let the plant bounce back and forth between high and low alarms."

What a gulf between management perception and the reality of the control room! The managers and engineers, who had never hung out in the control room all night long, were totally unaware of the risks they faced with foggy-headed operators potentially making unfortunate decisions with explosive chemicals and multimillion-dollar equipment. The operators were lulled by the false security of the high-tech equipment; it is hard to get excited when nothing much happens night after night. Yet their job is best described as in the category of 99.9 percent boredom and 0.1 percent terror. Danger is always lurking, things can always go awry. In the absence of fear, the crew sought comfort for those long nights in the control room. And in the middle of the night, seeking comfort entails making adjustments that reduce your level of alertness and risk putting yourself to sleep.

THE MISSING ELEMENT IN MANAGING HUMAN PERFORMANCE

THE MANAGERS and engineers who wrote the job description for those control room operators did not consider the design specs of the human body during the nighttime hours. The plant's equipment was designed and management systems set up during daytime hours by people who were not constantly reminded about fatigue by working shifts around the clock. They assumed that the operators would stay alert, but they had created a work environment in which the operators were bound to fail.

These managers and engineers were not alone. Classically, managers and engineers have been taught to use methods based primarily on psychology and ergonomics to ensure the optimal performance of their employees. This traditional approach relies on paying attention to three key factors (Figure 4.1):

1. Aptitude and Education
Recruit and select employees with the best aptitude, education, and skills required for the job.

2. Training and Experience
Provide the most effective training courses and on-the-job training, or make sure employees have already received it, so that they fully understand their job description, goals, and targets and can perform the necessary tasks to perfection.

3. Attentiveness
Try to ensure that employees pay full attention to the tasks at hand. Applying the principles of psychology and ergonomics, seek to address the following:

a. Motivation. Devise the most effective ways to motivate employees to perform at their best. Managers constantly develop incentives to ensure that employees pay full and energetic attention to their jobs. These may be simple carrot-and-stick approaches or more effective self-motivation strategies using participative management techniques.

b. Work Load. Design jobs using ergonomics and automation to ensure that employees are not physically or mentally overloaded, which impairs their ability to perform. A nuclear power plant operator in an emergency may be faced with panel after panel of flashing lights, hundreds of meter readings that may be inaccurate, and bursts of excited voices transmitted by the walkie-talkie radios car-

Figure 4.1

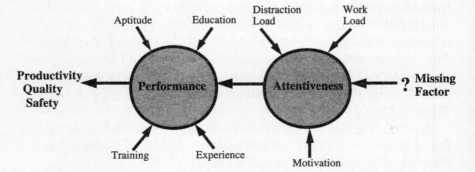

ried by operators out on the floor of the plant. Engineers and managers are vigorously addressing such problems by applying ergonomics to job design and relieving humans of many of the tasks they used to perform.

c. Distraction Load. Reduce the number of stimuli in the work environment that are not essential to the tasks employees perform. Such distraction loads have an important and well-understood influence on performance. A student cannot do a difficult homework assignment with the TV on; a pilot should not discuss a real estate deal while bringing the plane in for a landing at Los Angeles International Airport. This factor receives significant attention from managers and engineers. Indeed, engineering solutions often pay considerable attention to reducing extraneous noise, visual cues, and increasing the comfort of the workplace.

The myth is that a manager who has energetically addressed all of these factors will obtain optimal human performance. A supermanager will even throw in health programs, fitness centers, and culture-building activities as well. But a critical factor, especially important to manage in around-the-clock operations, is still missing.

THE MANAGEMENT OF ALERTNESS

TRADITIONAL APPROACHES to the management of human performance may be good enough for a nine-to-five world, but they are seriously inadequate when people are deprived of sleep and must work at all hours of day and night. The missing element in this equation of optimal human performance is the *alertness* of the person—a critical, and often overlooked, component of their attentiveness to the task at hand (Figure 4.2).

Alertness is the optimal activated state of the brain, when the fog of fatigue has lifted, when the brain hums and purrs, when creative solutions to old problems pop into the mind. Alertness

Figure 4.2

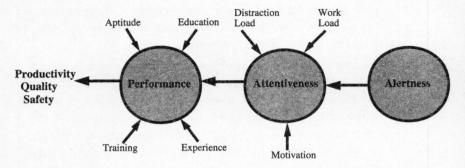

enables us to make conscious decisions about what to pay attention to in our environment and what to exclude. Alertness keeps us out of trouble, on the highway, in the cockpit, on the factory floor or the trading floor. Without alertness, there can be no attentiveness, and without attentiveness no performance. All the selection, training, and motivation in the world does no good unless the human brain is clearheaded and alert. Whenever our brains are fogged in by fatigue so that a thought can barely land, let alone be correctly processed, human performance can be catastrophically impaired.

The key difference between alertness and the other determinants of human performance is that alertness dynamically changes with time. Aptitude and education, training and experience, change only slowly across days and months, but alertness may vary from second to second and must always be watched. This means that the manager who seeks optimal human performance around the clock must pay special attention to moment-to-moment alertness, for this determines more than anything the performance of the human machine.

Alertness can be a complicating factor, because its management may require actions that sometimes run counter to traditional approaches to manage the other determinants of human performance. Take, for example, the management of distraction load and work load. Managers and engineers commonly think that both should be reduced to improve human performance. However, remove too many distractions and reduce the work load too far, and alertness will deteriorate as monotony sets in (Figure 4.3). With little to do in the middle of the night, a person

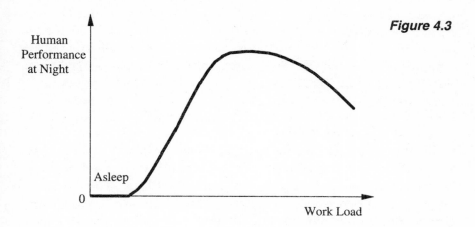

Figure 4.3

can too easily experience waves of sleep that reduce human performance to zero. That is why a car radio can be a godsend when one is driving alone and feeling sleepy, and that is why banning radios from an industrial control room when all is humming quietly in the middle of the night does not make sense.

THE INGREDIENTS OF ALERTNESS

THE SAME Professor Walter B. Cannon who taught us about homeostasis also clarified the functions of the autonomic nervous systems of the body that regulate all the essential functions "nature saw fit to remove from the caprice of an ignorant will." These nervous systems—the *sympathetic nervous system* and the *parasympathetic nervous system*—are very helpful in clarifying the concept and mechanisms of human alertness.

The sympathetic nervous system is the one automatically triggered in emergency situations. It makes the heart pound, the blood pressure rise, the pupils dilate, and the hair stand on end. This is what Cannon graphically described as the "fight or flight response." The body is placed on extreme alert, ready to make life-or-death decisions, to battle the enemy, or to flee from danger. Alertness is at its peak when the sympathetic nervous system is fully activated.

The parasympathetic nervous system is the counterbalancing

automatic (the correct medical term is *autonomic*) system. This is a second special system of nerves coursing throughout the body which switches a person into a relaxed state. Envision an old man dozing in front of the fire after a heavy meal. His parasympathetic nervous system is responsible for his constricted pupils, his slowed heart rate, his lowered blood pressure, and the borborygma (the onomatopoeic medical term for the rumbling) of his fully activated digestive system. Clearly, here alertness is at its lowest ebb.

Human alertness is best thought as the balance between the sympathetic and parasympathetic effects on the human brain. This hypothesis is intriguing and helpful, and the data are pouring in to support it. It enables us to design effective strategies to manipulate the level of a person's alertness—to flick the switches of this automatic system at our own choosing. Nine key switches are available, but before discussing them further we will examine how to measure the level of human alertness.

TAKING THE MEASURE OF ALERTNESS

HOW ALERT are you now as you read this page? Some measure can clearly be obtained from subjective evaluation. You can tell me that your alertness fluctuates and can indicate where you are now on a scale such as the 7-point Stanford Sleepiness Scale. This provides an easy way to keep track with scores varying from 1 ("feeling active and vital, alert and wide-awake") to 7 ("almost in reverie; sleep onset soon; lost struggle to stay awake"). However, this simple technique is heavily biased by what a person wants others to believe. Can you see nuclear power operators voluntarily recording their true levels of sleepiness for their supervisor, when drowsiness at the controls is cause for disciplinary action?

Another age-old technique is to station an observer to watch for the characteristic behavioral changes of drowsiness. Eyelids droop, eyes are bloodshot, blinking increases, yawning becomes uncontrolled, the face loses expression, speech is slowed—these are the well-known giveaway signs that a person is losing the battle to stay awake. Unfortunately, the reliability of such data is impeded by the infectiousness of fatigue. Nothing more quickly puts one to sleep than having to watch another person struggle with the waves of sleep. Once the observer's attention lapses, the reliability of the data obtained is destroyed.

With the advent of techniques to record brain electrical activity, it has become possible to use special equipment to keep track of the state of brain activation. By measuring the brain waves with an electroencephalogram (EEG) and eye movements with an electroculogram (EOG) from electrode wires attached to the scalp, plugged into a portable Sony Walkman-size recorder, we can track, second by second, the sleepiness and alertness state of people as they go about their daily business and rest. It is rather like wearing a Walkman in reverse—the information in the wires travels from the head to the recorder and is collected on the audiotape, rather than being dispensed from it and being sent to one's brain.

When alertness becomes significantly reduced, changes start to occur in the waking EEG. The brain waves show brief snatches of patterns normally seen during the stages of sleep. Subjective and behavioral sleepiness are associated with increased alpha waves (8–12 Hz) and increased theta (3–7 Hz) activity. After an approximately fivefold increase in alpha or a twofold increase in theta, people can no longer interact effectively with their environment. At the same time, slow rolling eye movements may occur and the pupils may become constricted.

Electrical episodes normally associated with sleep may briefly intrude into the waking EEG. Hiccups in the brain waves, called K-complexes, may be seen even though they are normally seen only in stage 2 sleep. Then bouts of deep delta waves may appear. These episodes, called "microsleeps," resemble the slow waves normally seen only during the deepest stages of sleep. Usually quite short, intruding for only ten or fifteen seconds or so, they are nevertheless scary when a person is performing a critical activity such as driving a car or flying a plane.

The standard laboratory test used to assess alertness today is the multiple sleep latency test (MSLT). This test, although simple in concept, is laborious to perform. The subject lies on a bed in a darkened room, and researchers, using brain EEG recording electrodes, determine the time the subject takes to fall asleep. A person with very low alertness will drop off in a minute or two, whereas an alert person will spend the entire allotted bed time of twenty minutes wide awake (Figure 4.4). Purists will argue that this test is really one of sleep tendency, but it provides a pretty good sense of an individual's alertness level.

The MSLT's greatest value comes from the extensive data that scientific researchers have accumulated on the effects of all the factors that influence alertness. The MSLT enables one to put numbers on the relative effects of the loss of two hours versus four hours of sleep (Figure 4.5), or on the effect of time of day and caffeine consumption, for example. Hence the MSLT serves as a benchmark for evaluating factors that influence alertness. We have also used it, as discussed in Chapter 11, as a means of reconstructing events in legally contested cases in which human alertness is at question.

But the MSLT is a laboratory test—it is of little or no use in the real world. So the technology of human alertness measure-

Figure 4.4

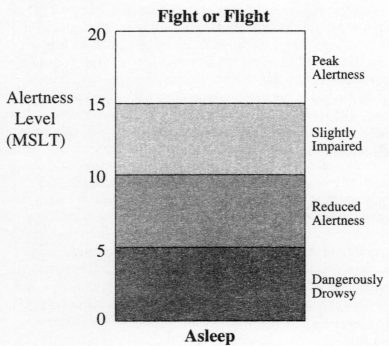

Figure 4.5
Effect of Reduced Sleep Time on Alertness

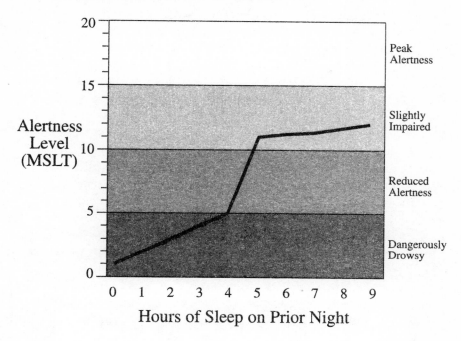

ment is currently racing toward the development of real-time, fast, accurate, and responsive tests that can be used to assess the activation state of the brain. To make full use of such measurements as they become available, we will need to know how to switch alertness on and off at our command.

THE NINE SWITCHES
OF HUMAN ALERTNESS

A PERSON'S alertness is triggered by nine key internal and external factors that can be considered the switches on the control panel of the mind. Understanding these key switches and how to manipulate them is the secret of gaining power over one of the most important attributes of the human brain (Figure 4.6).

Figure 4.6
The nine switches of human alertness.

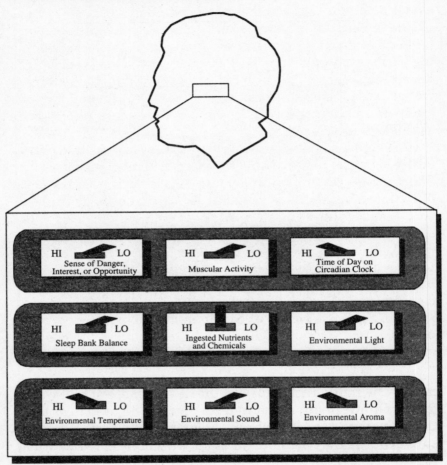

• Switch 1: Sense of Danger, Interest, or Opportunity

Nothing pulls us faster from a drowsy state than the imminent threat of danger, or just surviving a near miss. The emergency flight-or-fight response is activated by the sympathetic nervous system, and the brain is placed on full alert. The heart pounds, the hair stands on end, and the skin turns cold and clammy.

The presence of danger is not enough, however; the danger must be perceived and feared. The swimmer gobbled up by a shark from behind may not trigger the response until too late.

An inadequately trained operator in an industrial plant may not recognize the impending danger of an explosion, and an operator who is asleep at the switch certainly will not.

The farther one is from perceiving danger, the less the sympathetic system is triggered from this source, the more the relaxing parasympathetic tone is allowed to reign. One is lulled, falsely or not, into a sense of security. Maintaining a certain controlled level of anxiety—or at least concern—is helpful for optimal performance whenever a person is working in what is called a "safety-sensitive job."

Although less extreme than the response to danger, a stimulating job, task, or opportunity triggers a similar response. Watching the lottery balls plop into their slots is a boring piece of TV trivia unless one is holding a ticket and already three of your numbers have come up. The pupils dilate, the heart rate starts increasing, and the hopeful ticketholder becomes animated. The sympathetic system is triggered and the parasympathetic suppressed, by this as well as other stimuli such as an interesting challenge at work, an exciting idea, and anything else that is new and different.

But if the job is boring or monotonous, the alertness of the person fades. The same old thing, the too familiar, the endless stretch of freeway across the western desert, the night shift in a plant when everything is running smoothly all trigger parasympathetic doziness. When alertness is lost, danger looms, for attentiveness and performance will also suffer.

• Switch 2: Muscular Activity

Any type of muscular activity triggers the sympathetic nervous system and helps you to keep alert. You do not have to be running a mile or lifting weights—taking a walk, stretching, or even chewing gum can stimulate your level of alertness. After vigorous activity the effect can last an hour or more; it is very hard to fall asleep immediately after returning from a jog.

The trouble is that many of the most dangerous activities in our technological society require us to be sedentary. We sit for hours on end while driving our cars, air-traffic controllers and nuclear power plant operators sit in front of their control consoles, and pilots sit in the cockpits of their airplanes. Just when we most need to be alert, we instead reduce our level of muscular activity and trigger parasympathetic relaxation.

• Switch 3: Time of Day on the Circadian Clock

The self-same mechanisms of sympathetic and parasympathetic control are used by the biological clock as it wakes us up ready for the day ahead, and then settles us back down toward sleep as the day fades away. The giveaway signs of sympathetic activation are there to see in the wake-up reveille. Metabolism is activated, skin blood flow shut off as the body warms, and heart rate and blood pressure start to rise—all evidence for the role of sympathetic tone. Then, as drowsiness sets in in late evening, the pupils constrict and the body temperature falls as parasympathetic mechanisms prepare us for restful sleep during the night ahead. Our alertness cycles systematically and logically throughout the day (Figure 4.7).

Figure 4.7

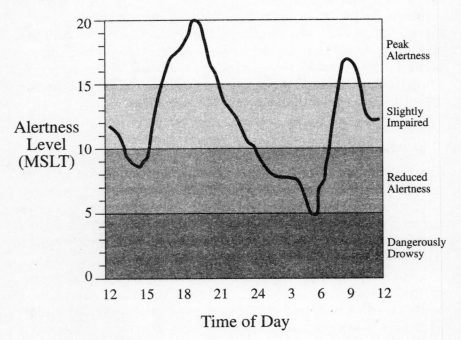

These mechanisms, so perfectly attuned for the traditional pattern of daytime wakefulness and nighttime sleep, get us into trouble in our twenty-four-hour world. The biological clock, and therefore the coordinated cycle of parasympathetic versus sympathetic tone, is slow to adjust its timing to any newly imposed schedule. Alertness may be triggered when it is time for sleep, and sleepiness triggered when we need to stay awake.

However, the circadian clock can adjust slowly to successive nights of shift work and daytime sleep. Thus the drop in alertness during the early morning hours is most marked on the first two nights on shift, and thereafter starts to adapt. Within a few days, if light and dark cues are controlled, alertness is more easily sustained at night and daytime sleep improves (Figure 4.8). The problem is, when you come off the night shift or take a few days off, the clock has to be reset all over again. Part III covers new technologies that provide solutions to this problem.

Figure 4.8

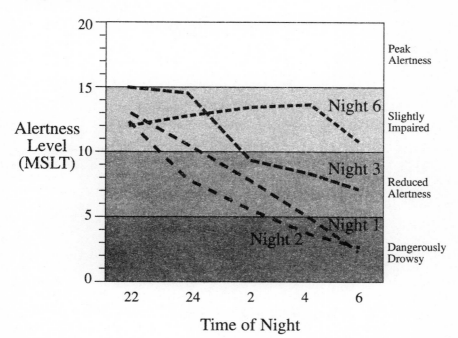

• Switch 4: Sleep Bank Balance

Aside from the modulating influence of the biological clock, sleep tendency has its own agenda, driven by the number of hours since we last slept. Recuperative sleep makes deposits in our "sleep bank," and sustained wakefulness makes withdrawals. When you have withdrawn too much and the sleep bank balance is too low, the pressure to sleep may become extreme, especially after twenty-four hours of sustained wakefulness, whatever the time of day. This will have a severe dampening effect on the alertness of the brain (Figure 4.9).

Such drastic sleep deprivation is not necessary for effects on alertness to be observed. Lopping a couple of hours off one's regular sleep length for several consecutive nights will result in a steady fall in alertness levels. A sleep debt will be accumulated and will be manifest in the level of daytime alertness (Figure 4.10). Similarly, reduced and disrupted sleep during the day severely reduces one's level of alertness on the night shift.

Figure 4.9

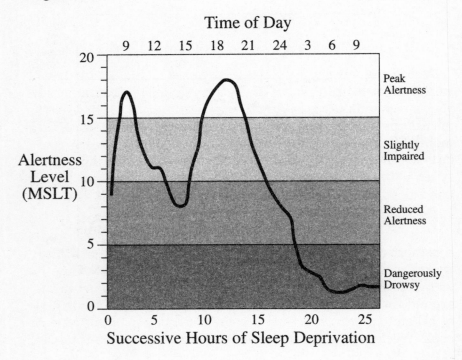

Time of Day

Alertness Level (MSLT)

Peak Alertness

Slightly Impaired

Reduced Alertness

Dangerously Drowsy

Successive Hours of Sleep Deprivation

Figure 4.10

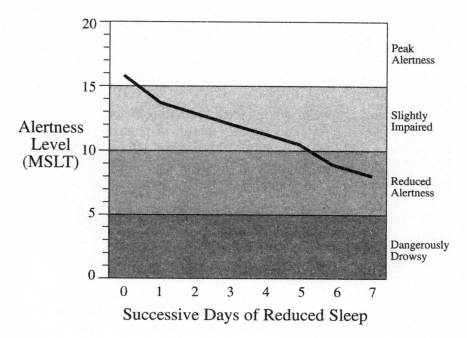

A good single night's sleep unrestrained by the constraints of worldly time schedules can do much to replenish the balance in your sleep bank account, as can naps strategically scheduled for maximum effectiveness. In older people, sleep may become fragmented or awakening may occur too early in the morning because the aging biological clock has a tendency to slip out of synch with the day-night cycle. This may result in some accumulated sleep debt and therefore reduced alertness.

• Switch 5: Ingested Nutrients and Chemicals

Alertness may be altered by the chemicals and nutrients that we ingest or inhale. This is often a maladaptive way to force alertness on a fatigued brain, but is resorted to by many in the around-the-clock operations of our society. Coffee drinking may be taken to an extreme—shift workers drink pots of coffee, not cups. Truckers ingest amphetamines to keep going on the road, and, in earlier, more innocent days, hospital interns and others trying to stay awake on the job relied on small sniffs of cocaine.

Sometimes to reverse the sleep-destroying effect of uppers, millions of insomniacs swallow millions of sleeping pills in an effort try to control sleep. This is big business for the pharmaceutical industry, even though the effectiveness of such drugs is limited.

One of the biggest problems with pharmacological strategies is that the brain chemistry rapidly adapts so that larger and larger doses are needed to obtain the desired effect. The insomnia rebound is also troublesome as one tries to wean oneself from these aids. Some of the alertness-inducing or -suppressing effects of these drugs may be mediated by adjusting parasympathetic and sympathetic tone, but other chemical changes in the brain undoubtedly also play a role.

• Switch 6: Environmental Light

Restaurateurs have played with the effects of lighting for years. Fast-food places are brightly lit, encouraging fast eating and fast departure, whereas in expensive restaurants, where one lingers and pays appropriate parasympathetic homage to a gourmet meal, the lights are turned down low and the pace is slow.

Bright lights—1,000 lux or more—can have dramatic effects on suppressing sleepiness especially on the night shift and keeping the staff in a state of sympathetic activation (Figure 4.11). Attentiveness and performance shoot up and the risk of errors is reduced. We will return to this important technology in later chapters.

• Switch 7: Environmental Temperature and Humidity

It is common experience that cool dry air, especially on the face, alerts one out of a sleepy state and that sultry heat brings on the desire for sleep. Similarly, a cold shower may be invigorating, a warm bath preparative for one's beauty sleep. These activities trigger the appropriate sympathetic or parasympathetic response, as blood is shunted to or from the skin and the brain in parallel is alerted or slowed down.

Figure 4.11

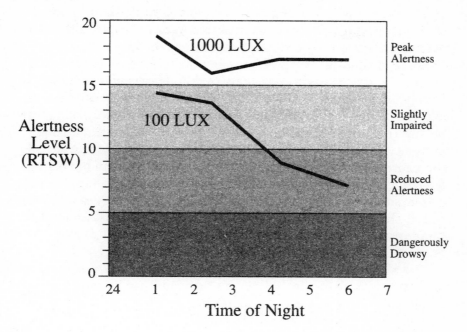

• Switch 8: Environmental Sound

Sound can invigorate us or send us to sleep. The rolling surf on the beach, the smooth rushing of a mountain stream, can lull us to sleep—so effectively that these sounds are now electronically simulated in "white noise" machines that many people use in their bedrooms. Unfortunately, the same electronic white noise is produced in less desirable places, such as industrial control rooms, by the equipment that people are meant to watch alertly through the night. In contrast, irregular or variable sound will help keep you awake—such as the intermittent creaking of a door gusted by drafts in the middle of the night.

• Switch 9: Environmental Aroma

Although less scientific work has been done on aroma than on the other eight switches, reports and claims are starting to assert that some aromas, such as peppermint, make people more alert. Whole technologies are being developed—Shimizu in Japan is playing a leading role—to deliver wafts of aromas through specially designed ventilation systems to have the stimulating effects on the occupants of the room.

Some of these aroma effects may also be triggering switch 5— ingested nutrients and chemicals. At the Nestlé coffee factory in Ohio, which serves as the central location for the decaffeination of coffee beans for Nestlé facilities around the world, an incredible array of sacks of coffee creates an environment of delicious aromas. More to the point, sacks of caffeine dust extracted from the beans are permeating the atmosphere. Does this have an effect on alertness? In every other shiftwork plant I visited, they drink gallons of coffee around the clock to try to stay awake. At Nestlé, however, there was not a coffee pot to be seen, and fewer complaints of sleepiness could be heard. Maybe the workers there inhale all the caffeine they need.

The nine major switches of human alertness act in concert to modify our alertness level hour by hour, minute by minute. Understanding the routes to achieving an alert optimally performing brain, running smoothly on all cylinders, gives us the potential for control—the ability to pull the levers of alertness.

COMFORT VERSUS ALERTNESS

IT IS TEMPTING to say "Elementary, my dear Watson," because these truths may seem obvious and straightforward. However, we have not learned them as a society, nor have we used them even when we know them well.

One challenge is that the human desire for comfort intervenes. Making oneself comfortable is not compatible with optimal alertness, especially during the wee hours of the morning. In fact, the desire for comfort may be so dominant, and lack of awareness of the compromise one is making with alertness so large, that alertness takes a back seat.

In the high-tech industrial control rooms that are being installed all over the world—wherever there is a capital-intensive continuous process that can be automated—the lesson has not yet been learned. Control rooms, such as the one described at the beginning of this chapter, are more focused on human comfort than on alertness. In which direction had the nine alertness-influencing switches been flipped for that night-shift crew? The tally is straightforward:

Alertness-Influencing Factors	*Alertness Switch*
1. Sense of Danger, Interest, or Opportunity	OFF
2. Muscular Activity	OFF
3. Time of Day on Circadian Clock	OFF
4. Sleep Bank Balance	OFF
5. Ingested Nutrients and Chemicals	OFF
6. Environmental Light	OFF
7. Environmental Temperature and Humidity	OFF
8. Environmental Sound	OFF
9. Environmental Aroma	OFF

The pendulum was swung way over to the parasympathetic in that control room at night.

This case is by no means unique. Time and time again we come across the same situation in air traffic control, military installations, chemical plants, and oil refineries. Of even more concern, engineers are striving to build more and more comfort into the latest planes, process control panels, truck cabs, and railroad engines, in the mistaken belief that comfort improves performance.

The truth is, to be fully alert one must be a little uncomfortable. This is a hard, but important, lesson. We are not technologically ready for the lights-out factory and the pilotless plane—but because of inadequacies in engineering design and management, we have them by default.

Chapter 5

The
Costs
of Human
Breakdown

THE HUGE OIL tanker nosing its way out into
Prince William Sound late in the evening of
March 23, 1989, was carrying 1.3 million barrels of Alaskan
North Slope crude in its cavernous holds. Such ships need
human guidance to steer their way through the reefs and
islands of the Gulf of Alaska. Unfortunately, the humans on
board the *Exxon Valdez* that night were not up to the task. The
deck officers, exhausted from long hours on duty supervising the
loading of the oil cargo into the hold, were too fatigued to be sufficiently attentive to their navigation duties or adequately supervise the junior person placed at the helm.

Shortly after midnight, the *Exxon Valdez,* way off course, ran
aground on Bligh Reef and spilled more than eleven million gallons of crude oil into Prince William Sound. Twelve hundred
miles of coastline were polluted, 350,000 seabirds died, and
Exxon ran into a litigation nightmare the likes of which had
never been seen before. A $1 billion fine and $2 billion cleanup
fee have already been paid, and, at the time of writing, over $50
billion in outstanding civil suits remain to be adjudicated.

The captain eventually resorted to alcohol, but a jury concluded that he didn't start drinking until after the wreck had
happened that night. The underlying cause of the accident,
according to the National Transportation Safety Board, was
human fatigue.

64

WE CREATED our twenty-four-hour society, in part, as a way to cut costs, a way to squeeze more output from our scarce capital resources. Even the most rudimentary cost-benefit analysis shows that it is far cheaper to build one facility or one assembly line and operate it twenty-four hours a day, seven days a week, than to build four facilities or assembly lines and operate them each only forty hours a week. More product can be manufactured in that single continuous operation, because daily start-ups and shutdowns are eliminated, and the savings on the capital costs of the equipment are enormous, particularly if by turning your facility or assembly line into a continuous facility you increase the opportunity for automation and thus reduce the need for personnel.

Thousands of companies around the world are realizing the benefits of such savings in capital costs. Older factories are closed down, while newer facilities, or the ones more amenable to automation, are expanded in capacity to replace their output, not necessarily by investing in more equipment but by expanding the work hours around the clock and around the week, so that in a smoothly running operation 10:00 A.M. on Tuesday is no different from 3:00 A.M. Sunday. This transition is gathering such momentum that our consulting group, Circadian Technologies, has a thriving operation devoted just to converting plants from five-day operations to seven-day operations around the clock.

Many businesses today just would not make sense operated from nine to five. Exxon or Mobil could not be expected to anchor their oil tankers at night and during the weekends and sail only during weekday daytime hours to ensure a fully rested crew. Similarly, no one would want medical care, electric power, police, and security on less than a twenty-four-hour basis. We have grown to rely on these services, on being able to telephone anybody at any time, to find that doctor or drugstore at 3:00 A.M. when a child is seriously ill. We do not want to waste time traveling by confining all trips to daytime hours. The pace of life is too demanding, even though we must therefore rely on pilots and air traffic controllers who live on disrupted sleep-wake schedules to service our needs. In tough economic times, we also rely on the cheaper prices for goods and services that businesses can offer by cutting their capital costs through around-the-clock operations.

The benefits of our twenty-four-hour society are enormous. This is a tide that cannot be turned. But the costs are also enormous—especially if we do not master the art and science of living in a nonstop world. It is essential that we identify the sources of these costs so we can control them. But this requires an understanding of how people fail when fatigued.

THE COSTS OF ACCIDENTS

THE COSTS of an individual incident of human inattention can vary greatly. When my attorney nodded off to sleep at the wheel late at night after a long day in the office, he was at slow speed on a side street and all he hit was a telephone pole. The cost was one or two thousand dollars of damage to his car and a bruised ego. At the other extreme, the total cost of the *Valdez* incident, involving just one hour's worth of human inattention, will eventually total many billions of dollars for Exxon, even if only a fraction of the outstanding legal claims are awarded. Such sums would sink many another, smaller company, and there are much smaller companies than Exxon that operate oil tankers.

Even greater costs were incurred in the Chernobyl nuclear power plant disaster, caused directly, as shown in Chapter 8, by operator carelessness and inattention in the wee hours of the morning. At the latest count, 300 people were killed, 1.5 million others were contaminated, and hundreds of square miles of countryside were made unfit for use. The total cost of cleaning up contaminated areas, resettling people, and providing medical care and uncontaminated food for huge sections of the population has been estimated at more than $300 billion, even before the effects of post–Soviet Union inflation are taken into account. Even if one takes such estimates with a pinch of salt, one cannot dismiss the $26 billion appropriation that has already been spent by the Supreme Soviet for emergency costs alone.

And we have not touched on Three Mile Island, a notorious nighttime accident involving human error, nor the major chemical spill into the Rhine, which heavily polluted that central waterway through Europe. Each cost billions of dollars in damage.

Other costs add up to large sums because of multiple less costly incidents, such as the $9 billion per year cost of highway accidents incurred by the U.S. trucking industry. Since driver fatigue has been determined as the major cause of over half the truck accidents in the U.S., we can assign approximately $5 billion as a fatigue-related cost. That cost turns into higher insurance premiums, and in turn higher operating costs, and higher prices for shipping goods for everyone.

Individual major airliner crashes cost approximately $500 million, according to the U.S. National Transportation Safety Board, and even with the high level of safety we may expect several such accidents a year. Since two-thirds of aviation accidents are due to human error and many are related to fatigue (see Chapter 7), the costs here also stack up.

Our surveys of industrial plants consistently show that shift workers have twice the number of highway accidents as day workers. After struggling to stay awake all night at work, they too often cannot keep their eyes open on the drive home. Considering that the National Safety Council puts the annual cost of highway accidents in the United States at $49 billion and 10 percent of the working population work nights or rotating shifts (and after correcting for the percentage of the driving-age population who are not in the work force), we can calculate that the costs from this one highly specific cause amount to $2 billion or $3 billion per year in the United States alone.

Consider deaths and injuries at work caused by industrial accidents. The National Safety Council estimates that such deaths and injuries cost the U.S. economy $37 billion per year. We know that errors of inattention, which are the ultimate cause of many such accidents, are doubled or tripled at night and on rotating schedules. Taking the percentage of such work into account, we can determine that $1 billion or $2 billion more can be laid at the feet of human fatigue in twenty-four-hour operations.

What is the total cost of accidents caused by human fatigue in a nonstop world? Most of the numbers used in these calculations are taken from statistics on the U.S. economy. To estimate costs for the world, I use the fact that the U.S. economy represents about 20 percent of the world economy and therefore multiply the estimates accordingly. I have assigned the cost of the annual $10 billion major catastrophe to the world, although from time to time the United States is the victim, as with the *Exxon Valdez* incident. Here is the estimated tally:

Cost of Human Fatigue–Caused Accidents	**U.S. Total Cost** ($ billion)	**World Total Cost** ($ billion)
Major catastrophes (*Valdez,* Chernobyl, Bhopal)	2	10
Lesser catastrophes (airline crashes, major plant explosions)	5	25
Industrial accident death and injuries	1.5	7.5
Truck accidents	5	25
Night-shift worker automobile accidents	2.5	12.5
Total accident costs caused by human fatigue per year	*16*	*80*

PRODUCTIVITY AND QUALITY COSTS

IN THE MIDDLE of the night, the two vast paper machines cranked out the endless wide river of paper at speeds exceeding thirty miles per hour. Bleary-eyed operators watched for potential breaks, for keeping the machines continuously up and running meant the maximum productivity was obtained from the mill. Suddenly the paper snagged and tore on machine number two, and the lead operator reached over to switch off the machine so the tear could be fixed. The plant fell unusually quiet, and he realized that in his fatigue he had switched off machine number one instead of number two. He had shut down the entire paper mill and lost precious minutes of production because of the inattention of his tired mind.

FATIGUED PEOPLE make errors, most of them minor, but because there are so many they create an enormous effect. Fatigued people also work more slowly and less effectively. They do things the long and routine way, and fail to see efficient shortcuts that could be used. They do not pay attention: vats boil over, tanks overfill, tools drop into machines, and goods drop off vehicles. They inadvertently catnap, extend their meal breaks, and let their minds wander from the task at hand.

The manufacturing sector of the U.S. economy contributes $1 trillion in value added to raw materials each year. Thus even small changes in productivity that can be attributed to fatigue have considerable effect on the bottom line of our cost-benefit analysis.

The effects of fatigue, however, are not small, especially since most manufacturing in the United States is now undertaken in around-the-clock operations. Refrigerators, potato chips, semiconductor chips, diapers, ice cream, automobiles, tires, light bulbs, cameras—virtually everything—are manufactured around the clock. All of it is therefore highly susceptible to the effects of fatigue.

The more human intervention there is, the more the fatigue cost. Even if we are extremely conservative, there must be at least a 5 percent fatigue-induced effect on productivity across the manufacturing plants of the United States, and probably much more. We have found that in consulting projects where fatigue in shiftwork operators has been reduced, the productivity of the operators is increased by 5 to 20 percent. If we take a conservative 5 percent figure and apply it to the $1,000 billion in value added per year to the U.S. economy, we obtain a $50 billion per year cost of human fatigue. Other related costs add up. Consider the downtime caused by industrial accidents and catastrophes, and also the effects of regulatory actions by federal agencies and litigation that saps the energy out of the corporate leadership of the company and therefore the economy.

Consider the example of the Peach Bottom nuclear power plant incident, to be discussed in more detail in Chapter 8. After a Nuclear Regulatory Commission inspector found some control room operators asleep on the job, the plant was ordered shut down and did not reopen for two years. No accident or injury had occurred, but it cost the company $14 million per month in replacement power it had to purchase (over $300 million in total) plus considerable recruitment costs to replace the echelon of management who were relieved of their posts, including the chief executive and the chairman of the board. Also consider the retraining and recommissioning costs, and the cost of defending against and settling a major shareholders' suit against the top executives.

Such impacts of regulation and litigation add up to enormous figures. In litigious societies like the United States, tremendous sums are awarded in the courts, of which the legal profession consumes approximately one-third. Even setting aside the amounts included as part of accident costs (see table below), several billion dollars more per year must result from these regulatory and legal issues and the lost productivity they mean for industrial operations.

Also take into account the consequences of employee turnover because of the inability of people to withstand shift work. There is a steady dropout rate of perhaps 1 to 2 percent per year. Each time a shift worker drops out, hiring and retraining costs are incurred to replace him or her. If it costs $10,000 per person on average to hire and fully retrain up to the competence level of the lost employee, and there is a 1 percent dropout rate among the ten million shiftworkers in the U.S. economy, then a cost of $1 billion per year can be assigned to this effect.

Again, we attempt to reach an order of magnitude number, with the same cautious qualifiers as before:

Lost Productivity from Human Fatigue	**U.S.** *Total Cost* ($ billion)	**World** *Total Cost* ($ billion)
5% reduction in manufacturing productivity	50	250
Productivity cost of regulatory and legal activity	4	12*
Employee rehiring and retraining costs	1	5
Total costs of lost productivity from human fatigue	*55*	*267*

*Assuming legal costs in the rest of the world are half of U.S. legal costs.

HEALTH CARE COSTS

PEOPLE WHO WORK in the around-the-clock industries of the twenty-four-hour society suffer a higher rate of illness and death than their counterparts on straight day shifts. Early studies seriously underestimated this effect by not recognizing that those who become ill may drop out of shift work and will therefore no longer be counted in the shift work population. However, when all exposed workers are included, the effects of shift work on cardiovascular disease and gastrointestinal disorders can be striking.

Anybody who has suffered from a severe case of jet lag understands the feeling of malaise, indigestion, fatigue, and insomnia that results from upsetting the timing of our bodies. But the problems of jet lag are usually minor and transitory compared to the difficulties faced by those who spend a lifetime working rotating or irregular shifts.

Not everyone is equally susceptible. In fact, it is becoming apparent that a subpopulation is particularly susceptible and will develop the full-blown condition we call *"shift maladaptation syndrome,"* (Figure 5.1) characterized by two or more of the following medical problems:

• Chronic Sleep Disorder

Virtually all who work rotating or irregular hours around the clock suffer some loss of sleep, but usually they recover well enough on days off and the problem does not get out of hand. When it does get out of hand, there is either an internal or an external reason for the severe and persistent insomnia. Internal reasons may be a biological clock that adapts poorly, or does not allow the adjustment of sleep patterns that are necessary for coping with changing schedules. The problem also may be precipitated by an underlying sleep disorder such as sleep apnea (cessation of breathing while in deep sleep) or narcolepsy (spontaneous lapses into sleep even when fully rested). External reasons may be a stressful marital situation or other crisis at home, or a very noisy home environment while the worker is trying to sleep during the day (Figure 5.2).

Figure 5.1

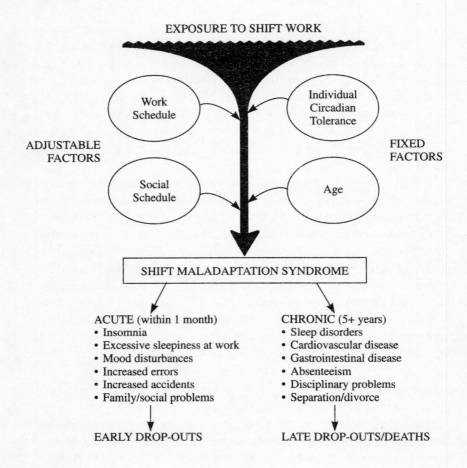

When such chronic insomnia occurs, a shift worker who has dragged herself through the night shift and has barely kept herself awake on the drive home may still have great difficulty sleeping when she gets home. This is particularly true on the first and second night shifts of the week, before her biological clock has had time to adjust. Three to five hours of broken interrupted sleep may be all she gets because the circadian body clock is chiming "Wake up—it's daytime."

The shift worker who drinks lots of coffee on the night shift to stay awake may make matters worse, for caffeine has a long

Figure 5.2

washout time before the body is rid of it and is likely to disrupt sleep several hours after the last cup.

When the worker returns to work the next night, her problems are compounded. She is starting out sleep-deprived and is still working on the backside of the clock. She may drink even more coffee to try to stay awake and have an even worse time falling asleep after the shift. In this way, she can enter a vicious cycle and experience a continuous state of chronic sleep deprivation.

People may work rotating shifts for many years in their twenties and thirties without suffering any problems. However, in the early forties, fundamental changes typically occur in the patterns of sleeping and a serious intolerance to rotating shifts may develop. The manager may see a highly reliable employee deteriorate into an exhausted, moody, and difficult individual, and may even fire the person if the problem is not recognized in time. With the right diagnosis, however, the problem is often readily manageable.

• Gastrointestinal Disorder

The digestive system has its own precise clockwork that times the secretion of gastrointestinal enzymes and hormones in anticipation of normal mealtimes. These are released even before sight or smell of food and account for the hunger pangs one gets at strange hours when traveling away from one's home time zone.

On the topsy-turvy schedules of a nonstop world, we deposit food in our stomachs at times when the digestive system is least expecting it, or we do not eat when the system is expecting nourishment. That pattern repeated over time causes stress, dyspepsia, and ulcers.

The incidence of ulcers tends to increase after about five years of exposure to repeatedly shifting schedules and mealtimes. It appears to account for a two- to eightfold increase in ulcer risk depending on what opportunities the person has to return to a normal, more ordered life. Even in the absence of a fully developed case of ulcers, the dyspepsia contributes to the general malaise that many shift workers experience.

Careful nutrition, and choosing what one eats and when one eats it, can help. This is hard to manage, however, in a lifestyle that includes breakfast at dusk and dinner at dawn.

• Cardiovascular Disorder

Studies of rural populations in Sweden where people stay in shiftwork jobs because they have few alternatives show two- to threefold increased risks of coronary artery disease and myocardial infarction (heart attacks) after five to ten years. We have also seen evidence for similar increased risk in overnight courier airline pilots, as compared to the general population of commercial aviation pilots.

To determine the full effect, such studies track all the people who drop out of shift work for any reason, as well as those who stay employed. These studies also have been designed to make sure that smoking and other related risks do not contribute. One major difference between shift workers and day workers appears to be increased lipid and cholesterol levels. This finding is supported by recent animal studies that show that cholesterol levels are elevated when animals are placed on rotating schedules.

• Mood Disorder

A common complaint from spouses and employers is that shift workers are like grumpy bears when on certain shifts. With shift maladaptation syndrome, the changes may be more severe.

Considerable evidence shows that mood is influenced by the precise setting of your biological clock with respect to the day-night or work-rest cycle. Malfunctions in that relationship can lead to a severe winter depression (seasonal affective disorders) when days are short. We suspect a similar chronic depressive state may occur in some individuals due to the disordered structure of internal time caused by shift work.

• Chronic Illness

Preexisting chronic diseases such as diabetes and epilepsy become much harder to control on irregular work and sleep schedules. Indeed, these should be treated as reasons to exclude an employee from a shift work job.

NOT EVERY person who develops shift maladaptation syndrome displays the full pattern of all these disorders. The common feature, however, appears to be the chronic sleep disorder, which is typically the first to develop and which may underlie the development of the other medical problems.

People most susceptible find they just cannot cope with shift work, and, if they can, soon quit for a day job. This constant self-selection process makes the incidence of shift maladaptation syndrome much less than it would otherwise be. Some sense of this can be gained from studies of wartime production plants in which people were put on rotating shifts and stayed there out of patriotic duty. The rate of stomach ulcers was eight times higher than their day-shift-working counterparts. However, in self-selected peacetime shift work operations, the rate of stomach ulcer is only twice as great as the nonshift-working population.

"Simple," you might say, "let's just keep all the people with a proclivity for shift maladaptation syndrome on day jobs." Unfortunately, it is not quite so easy. In many remote rural communities, by far the best-paid jobs are shift work jobs in local manufacturing plants. With overtime, workers may be taking home twice or three times what they would earn elsewhere—a strong incentive to stay, even in the face of developing medical problems.

The medical costs are significant for around-the-clock companies because most are actually or in effect self-insured. Cost estimates must be approximations and treated with due caution, and are subject to the normal interpretative cautions. For example, it is possible, although unlikely, that people who are drawn to shift work jobs are by intrinsic nature atypical in their susceptibility to disease and accidents as compared to the general population.

The largest cost results from the twofold or greater risk of coronary artery disease among around-the-clock shift workers. Since they represent about 10 percent of the work force and the total cost of coronary artery disease in the United States is approximately $47 billion per year, we can calculate that the extra cost is approximately $4 billion a year for the U.S. economy.

The costs resulting from the increased risk of gastrointestinal disorders, mood disorders, depression, and problems with managing other chronic illnesses also add up. Costs in terms of lost work days, and outpatient and inpatient medical costs, are less readily calculated but are considerable. My best estimate is $2 billion per year, leading to the following totals:

Increase in Health Care Costs	U.S. ($ billion)	World ($ billion)
Increased coronary artery disease	4	20
Increase in other health care costs	2	10
Total increased health care costs	6	30

SOCIETAL AND SECURITY COSTS

A NUMBER of long-term costs are even harder to quantify, but are real nonetheless. The most obvious are the costs of stresses on family life when one or both parents work around the clock or have irregular schedules. Just as people operating outside their design specs in a nonstop world may fail in their attentiveness to critical tasks, in their powers of reasoning and creativity, and in their health, so may they also fail in their social and family relationships.

• Failure of Parental Attention

Parents who work on certain shifts see little of their children. The evening-shift worker (4:00 P.M. to midnight, for example) becomes an absent parent because he or she arrives home after the family has gone to bed, sleeps late until after spouse and children have gone to work and school, and departs for work before the family comes back again. On fixed-shift systems, seniority arrangements may leave the worker with young children on the evening shift for years at a time.

It is becoming increasingly common for both parents to work shifts and turn over children at the factory gate. The problem with such child care is that the parent who is off shift has to sleep, and one cannot be much of a child care provider while asleep. This especially becomes a problem if a child is sick and needs special attention.

Another major trend sweeping the United States is the twenty-four-hour child care center, sometimes located at the workplace but more often at a remote facility. Extremely handy for the single parent, or two wage-earner shift work family, these arrangements are surely a stress and a deprivation of parental attention for the child who is dropped off at strange hours of the day and night and into a strange bed.

• Increased Family Stress

Reduced energy caused by fatigue and reduced time for communication because of irregular schedules can lead to enormous stresses in family life. Studies show that the divorce rate in shift working families is 60 percent higher than for day workers in regular jobs. As one operator put it, he had already in the past twelve years as a shift worker been through three wives, five houses, and six dogs! The stresses and strains on the family, the increased divorce rate, and the lack of attention by parents in a child's early formative years all can add up to very real costs for this generation and the next.

The security of nations also depends vitally on alertness and peak performance by military forces by night and day. The Persian Gulf War, shown or discussed in real-time twenty-four-hour a day by CNN, was an around-the-clock war. Fatigue was a major issue, as subsequent analyses and briefings show clearly. Mistakes by fatigued personnel can weigh in with heavy costs, especially in a nuclear age. Mistakes by tired leaders and decision makers trying to keep pace with the ceaseless flow of news and information can unleash devastating force, which may be initiated in error or without due thought when fatigue has set in.

We cannot expect to control these costs by retreating from the twenty-four-hour society into the lifestyle of the past. We are too far advanced in our technology and gain too many benefits from its application to our lives. Rather, we need to develop the wisdom to protect human needs and nature in the technological world we have created.

Total Costs of Human Failure in the Twenty-Four-Hour Society	**U.S.** ($ billions)	**World** ($ billions)
Accident costs	16	80
Productivity and quality costs	55	267
Health care costs	6	30
Societal and security costs	?	?
Total costs per year	*77+*	*377+*

As you can see, the costs are extraordinarily high, but the next chapters will show encouraging advances are being made in addressing these issues and reducing the risks and costs in many of the key industries of our society.

II

A
SOCIETY
IN
STRESS

Chapter 6

Aviation Safety
and
Pilot Error

THE BRITISH crew at the controls of the Lockheed 1011 Tristar had just crossed yet another five time zones after a flying schedule which, despite being legal, had not given them a full night's sleep for four consecutive nights. As they neared their next destination, at 3:30 A.M. home-base time, the captain overrode the autopilot so that the first officer could make a manual approach. Distracted by having to give directions on the position of the airport, the captain failed to notice that he was inadvertently holding the throttle in the idle position. At 3,000 feet, the flight engineer suddenly realized that the aircraft had slowed to *less* than the normal landing speed. They were only able to avoid a stall by the immediate application of power.

BY ANY measure commercial airlines offer remarkably safe transportation—much safer than driving one's car for the same trip. Advances in engineering and system management have brought huge improvements in the safety records of air travel. Yet it is the steadily reducing risk of equipment or system failure that has brought human error to the fore as a critical problem. Human error now causes 66 percent of all commercial airline accidents, 79 percent of commuter aircraft accidents, and a staggering 88 percent of all private aviation accidents.

The interlocked economy around the globe demands more and more international travel. The work and travel schedules of around-the-clock global business requires increased nighttime flying. Furthermore, because of engineering breakthroughs, the

81

distance planes can fly and the time crews must be on duty have steadily increased.

The limits of human tolerance in the aviation industry are being stretched to the hilt, and the prospects for reducing human error do not look good unless appropriate attention is paid to the needs of the human machine. There are serious restraints, because aviation is an extremely competitive industry, as the number of recently failed airlines shows. (Remember Eastern, Braniff, Pan Am, Midway, and People Express?) Profit margins are often thin, and the pressure to economize wherever possible is considerable.

Even in purely financial terms, however, attention to safety pays off. A major airliner accident can cost as much as $500 million, according to the National Transportation Safety Board (NTSB), and the long-term costs in loss of revenue can be high because a consistently poor safety record will rapidly cause loss of passenger confidence.

Not all the barriers to be surmounted require money, however. In fact, many require more of a mind-set change than a financial commitment, but that may make them no less hard to achieve. Until recently, for example, senior officials at the Federal Aviation Administration (FAA) consistently denied that aircrew sometimes fall asleep in the cockpit, in the face of overwhelming evidence from pilots to the contrary. In the face of such denials, it is hard to get regulations changed or progressive policies instigated.

WHEN THE HUMAN SUBSYSTEM FAILS

SEVERAL YEARS ago a second officer with a major airline submitted a list of fourteen incidents that had scared him so much that he asked to be relieved from nighttime flying. In addition to the incident at O'Hare that began this book, he tells of the morning in which his plane was lined up at sunrise to land on runway 6R at Los Angeles (LAX) when it had instead been cleared for an approach to runway 7L. Another time they responded inadvertently to an air traffic controller order to change direction that was meant for another plane, which itself was being piloted by another presumably sleepy crew that did not respond to the order. He said he could recollect a number of occasions when he found "the two people in front of me had nodded off at the same time, each expecting the other to stay alert." He himself was no more immune from sleep and recognized that he could report those incidents only when he himself "happened to be awake."

In another instance the entire crew of a Boeing 707 fell asleep in the cockpit while flying westward across the United States, and they failed to descend to their destination at Los Angeles. The plane, mindlessly guided by its autopilot, continued to fly at altitude out over the Pacific. The crew did not respond to voice directions from air traffic controllers, and for a while the situation looked hopeless. Finally, the controllers were able to trigger some chimes in the cockpit that woke the crew, and they got back to LAX before they ran out of fuel.

In 1990 a symposium on pilot fatigue at the Aerospace Medical Association attracted an overflowing audience to talk about a subject that until recently had been taboo. One pilot told of an airliner returning eastward across the Pacific toward Seattle which drifted into Canadian airspace and could not be contacted by controllers on the ground. Finally, they scrambled interceptor jets from the Canadian Air Force, which finally woke up the crew by shining bright lights into the cockpit. Another pilot told of a case just before touchdown, with the plane only 200 feet above ground, where the pilot fell asleep at the controls. Fortunately the co-pilot realized in time what had happened and was able to take over and land the Boeing 727.

While cruising at high altitude, the nighttime cockpit, in fact, provides conditions that are almost perfect for sleep. It is dark, because switching on the internal lights reduces one's ability to see oncoming traffic. One is strapped in a reasonably comfortable seat. Activity and exercise are not possible, and there is a severe letdown effect after the stress and concentrated work load associated with taxiing and departure.

In a study of night flights I undertook a number of years ago, I found dozens of incidents of nodding off in the cockpit at all times of day and night. But there was a tenfold increase in early hours of the morning, just when we would expect it, between

4:00 A.M. and 6:00 A.M. pilot domicile time, when the biological clock shuts down the brain. At this same time pilots flying flight simulators make the most errors (Figures 6.1a and b).

But nodding off, while the extreme, is not the only danger. Fatigue itself is a hazard to performance, and crews are often fatigued because of long hours, jet-lagged because of travel across time zones, and sleep deprived from having bedded down in strange hotel rooms at odd hours.

ACCIDENTS AND FATIGUE

WHILE a plane's equipment is wired into the black-box recorder that enables accidents to be reconstructed in sophisticated detail, pilots are not wired into the black box, other than through recordings of their voices or their manipulation of the controls. No recording is made of the human brain state so that no determination of the state of alertness or sleepiness, effective functioning, or zombielike behavior can be made. Hence it is very difficult in accident reconstructions to determine the precise nature of human failure.

Figure 6.1a

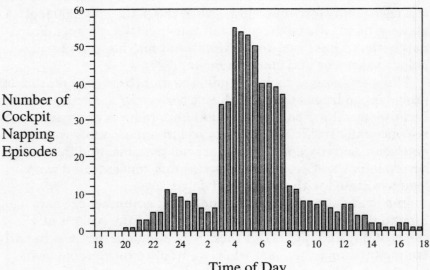

Number of Cockpit Napping Episodes

Time of Day

Figure 6.1b

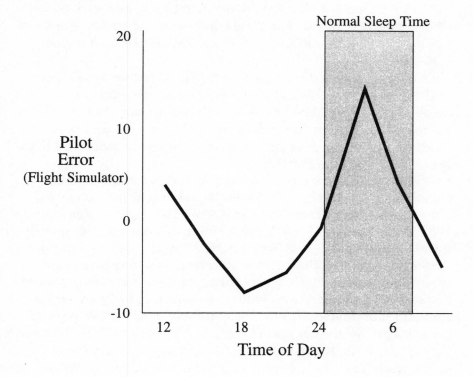

Fatigue particularly takes its toll in long overwater flights, especially over the Pacific. Crews are away from home for extended periods and often build up significant sleep deprivation plus chronic jet lag. Indeed, the FAA conducted a special investigation of Pan American's safety record over the Pacific from 1962 to 1974 because the airline's accident rate was 44 percent higher than the average of the entire domestic airline fleet. Looking in retrospect at individual accidents, the FAA determined that fatigue was an important factor. Exhausting routes flip-flopping between day- and nighttime flying were the norm. Clearly, such a schedule was a factor in the crash of a Boeing 707 at Bali in the Pacific in 1974, which killed all ninety-six passengers and eleven crew members.

The plane left San Franscisco at 7:44 P.M. and arrived at Honolulu at 1:32 A.M. The next day it left Honolulu at 3:39 A.M. and arrived in Sydney, Australia, at 2:35 P.M. The following day it left Sydney at 6:21 P.M. and arrived at Jakarta, Indonesia, at 1:30 A.M. It then left the same night at 2:18 A.M. and arrived in Hong Kong at 6:40 A.M. The fatal crash occurred in Bali the following day at 8:30 A.M. after the plane had left Hong Kong at 4:00 A.M.

Of course, such a duty-rest schedule does not necessarily precipitate an accident. Many crews follow such schedules even today (Figure 6.2). Most of the time they get away with it—there are many backups in the system. But when a crew is seriously fatigued, one of the most important back-stops in aviation safety is removed from the system.

NTSB investigators' conclusion that fatigue was a serious factor in an accident is not always included in the final report. For example, in a China Airlines Boeing 747 flight in February 1985, from Taiwan to Los Angeles, failure in one engine caused the air speed to slow and overwork the capacity of the autopilot to compensate. The tired and inattentive crew, ten hours into their flight and at 3:22 A.M., according to their body time, missed this critical problem. As a result, the plane suddenly entered a spin and plummeted 31,500 feet, seriously injuring two passengers, before the crew regained control. Although an NTSB investigator urged the board members to list fatigue as a probable cause, the NTSB chose to play down the role of fatigue. Subsequent analysis by an aviation human factors research team at NASA concluded inattention caused by crew fatigue was a key factor in this near disaster.

Analyses of other accidents have documented the important role played by pilot fatigue, including the KLM–Pan American collision in the Canary Islands in March 1977; the Eastern Airlines DC-9 crash in Charlotte, North Carolina, in 1974; the Pacific Southwest–Cessna collision over San Diego in 1978; and the commuter airline crash in Maine that killed the young Samantha Smith, well known and loved for her peace letters to Soviet leaders. Full recognition of the role of fatigue did not get into the final NTSB reports, presumably because of a mind-set block that simply assigned such factors to the blanket term "pilot error."

Fatigue, of course, is not confined only to the aircrew; ground maintenance crews can make fatigue-induced errors. After the

Figure 6.2
Irregular Sleep and Duty Times of Two Commercial Aircraft Pilots

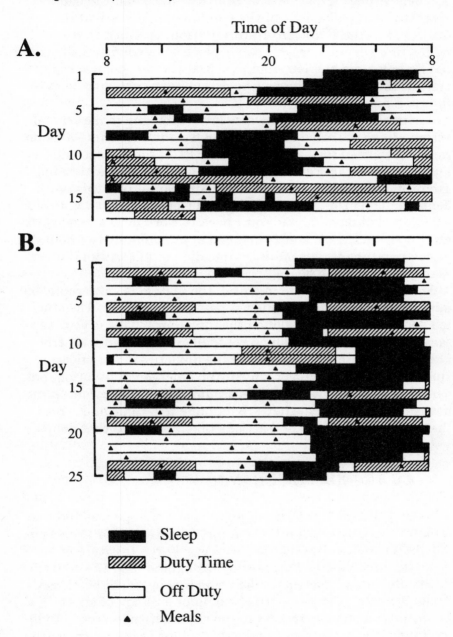

windscreen of a British Airways BAC 1–11 blew out as the plane climbed through 17,000 feet over Oxfordshire and the captain was partially sucked out of the window, the investigation, discussed in the Foreword to this edition, revealed that the maintenance engineer who had replaced the window using the wrong bolts, did the job from 3:00–5:00 A.M. on the first night shift of the week—just when he would be expected to be most fatigued and inattentive.

Air traffic controllers may also grow tired and make critical errors. When a United Airlines DC-8 freighter flew into a mountain in Utah at 1:38 A.M. in December 1977, some responsibility lay with the air traffic controller, who omitted a critical radial number in his holding instructions. After analyzing this accident, we found that the controller had worked a 7:00 A.M. to 3:00 P.M. shift that day and had come back to work for a second shift at 11:00 P.M.. The controller had had approximately two hours' prior sleep because his off-duty time did not fall within the normal nocturnal sleep hours.

Whenever humans are involved, the system is vulnerable, but admitting to human frailty is not part of the culture. The studied, laconic tone of the pilot on the intercom—"This is your captain speaking"—creates an aura of "all systems under control." One certainly would not want all the anxieties of the pilot shared with the passengers during an emergency situation, but denial can go too far. When it is assumed that aircrew are somehow superhuman, that they are immune to fatigue and stress, then systems designed and actions taken play into that denial.

AN AGENDA FOR CHANGE

ONE IMPORTANT first step toward change was the creation of the Confidential Human Incident Reporting Programme (CHIRP) which is run by the Royal Air Force Institute of Aviation Medicine at Farnborough, England, and the Aviation Safety Reporting System (ASRS) which is operated by NASA's Ames Research Center. With the help of a policy of strict anonymity, a steady stream of reports on altitude drops, flight path deviations, improper fuel calculations and landing on the wrong runway flow in to both CHIRP and ASRS each month.

About half of the reports that are submitted to CHIRP mention fatigue, lack of sleep or duty-rest scheduling issues.

Because of this overwhelming evidence, and because of the research done by pioneers such as Dr Tony Nicholson and his colleagues at Farnborough and the team at NASA, the inertia of the huge aviation bureaucracies is beginning to change. There are considerable challenges to be faced, however, in converting the mindsets of aviation engineers and managers so that pilot error becomes recognized as a fault of the system rather than just simply of the pilot. The key challenges are as follows:

1. to change the mind-set of aviation engineers so that man-machine interfaces in the airplane cockpit, and in air traffic control sector consoles, are designed with optimizing human alertness and effectiveness as one of the *primary* goals;

2. to change the mind-set of legislators and regulators so that practices that ensure aircrew alertness are allowed and promoted, and counterproductive regulations are banished from the books;

3. to convince airline management of the importance of improved sophistication of trip-planning algorithms, duty-rest scheduling practices, and personnel policies so that the effects of human fatigue are minimized; and

4. to train aircrew, air traffic controller, and maintenance personnel on techniques and strategies to maintain their optimal alertness and safe performance whenever required at any time of day or night. They ought to be, and for the most part are not, as knowledgeable about their own body and brain equipment as they are about the equipment they fly, direct, and maintain.

REDESIGNING THE MAN-MACHINE INTERFACE

THE AVIATION industry is remarkable in its technological achievements, yet that technology may force the human operator—the pilot or air traffic controller—to fail because of changes enforced by automation in the cockpit or air traffic control tower.

The aircraft cockpit is undergoing a comprehensive overhaul. Video monitors, computer keyboards, and precalculated routines increasingly replace the hands-on flying of previous eras. The Boeing 757s and 767s or the Airbus A300s are vastly different aircraft than the 707s and 727s. New generations of "glass cockpit" aircraft are in the works, providing more information than the pilot can ever use.

One driving concept in this next generation of aviation technology has been to reduce human work load, whether for the air traffic controller on the ground or the pilot aloft. Excess work load can certainly lead to human inattention and error, but a drive to reduce work load can go too far.

Just like operators monitoring an automated nuclear power plant or oil refinery, tired pilots with little to distract them in the middle of the night naturally tend to get sleepy or doze off. With the new generation of technology, the human operator risks becoming too much of a passive monitor—something the human machine is just not cut out to do when it is fatigued. The human sits there inactive, unchallenged, and bored. Then, when human competence has sunk to its nadir, automated systems may suddenly require human help when the system has malfunctioned—often posing a sudden challenge of complex problem solving and maximum work load.

Flying skills deteriorate when pilots rely too much on automation to do their work. They become reluctant to take over from the machine even when there is evidence of possible malfunction. The NASA ASRS reports show that specific automated devices, originally designed as backup systems to warn pilots— such as the altitude warning systems—have now become relied on as the primary source of information, after periods of reliable service. Thus the presence of automation has removed a level of safety protection because of the changed mind-sets of the pilots who use it.

These same problems are exemplified in the next generation of air traffic control equipment, called AERA-2. This mammoth, multibillion-dollar engineering project is named Automated En-Route Air Traffic Control–2 (one suspects they thought of the acronym first and then forced the full name into it!). AERA-2 combines the impressive technological prowess of IBM, Raytheon, and other centers of engineering excellence—and mindset.

Just read the description of this engineering marvel: it "introduces automation which reduces the controllers direct involvement in the formulation and issuance of air-traffic control clearances, and reduces the amount of conflict detection and communication and coordination with aircraft and other controllers." In other words, the air traffic controller is much less actively involved in making decisions, talking to pilots, or even other air traffic controllers. The machines with smooth efficiency take over many of the routine tasks that used to keep the controller awake and engaged.

The trouble is that an engineering project the size of the air traffic control system has a tremendous momentum of its own. Thousands of people are involved in its inception, and billions of dollars and many years are spent on its development. Although it is not realistic to call for a fundamental redesign at this stage, neither should the problem be ignored. The best approach would be a comprehensive audit of the systems, particularly the human interface, to see what environment is required for the human machine, and what work load and job demands are being created for the air traffic controller.

A similar comprehensive human alertness audit is needed for new airplane cockpit designs. In the air we expect a very high degree of human performance always to be available even if not always used, because of autopilots and cockpit automation. But the tests and trials of new aircraft examine in the closest detail the functioning of the machine and tend to overlook the impact of the design on the effectiveness of the human pilot.

The recently released National Plan for Aviation Human Factors from the FAA, which was spearheaded by Dr. Clay Foushee, is already having some impact on the industry with its recognition of the importance of human-centered automation. Also encouraging is the move by Boeing to hire to their cockpit design team a leading expert in pilot fatigue. Dr. Curt Graeber, who headed the research team on this subject at NASA Ames Research Center, has done much to build interest and an understanding of this problem in the aviation industry. I hope that his voice and influence will be heard within Boeing and that other major aircraft designers will follow suit in seeking such competent help. This is a subject foreign to the training and culture of most engineers, and as a result a subject that, if special steps are not taken, will tend to be ignored.

REGULATORY OVERHAUL

SOME OF THE most enlightened aircrew regulations in the world have recently been developed in the United Kingdom, thanks to the leadership of Air Commodore Tony Nicholson and his colleagues at the RAF Institute of Aviation Medicine at Farnborough. In 1990 the Civil Aviation Authority published "The Avoidance of Fatigue in Aircrew" (CAP 371) which is based on the latest research on human alertness and circadian physiology. The maximum permitted hours on duty depend on a sophisticated calculation based on several factors. They consider whether or not the aircrew has had time to adapt to a new time zone, how many time zones have been crossed, and what time of day it is. Furthermore, all airline staff who have responsibility for scheduling aircrew are required to receive training on the effects of circadian rhythm disruption and sleep deprivation.

In contrast, the American FAA regulations still equate eight hours of rest from 8:00 A.M. to 4:00 P.M., with eight hours of rest from 11:00 P.M. to 7:00 A.M., even though far more good quality sleep can be achieved during normal night-time hours than can be achieved during the day (unless you have just flown to the other side of the world). No allowance is made by the FAA for time-zone shifts or jet-lag and there is no recognition of the substantial scientific knowledge that exists on biological clocks and sleep.

For years scientists who do know about human sleep and circadian rhythms have been trying to persuade the FAA to modify their flight-time/duty-time regulations. They want them to be more in keeping with the huge body of scientific evidence (and common sense) and says that the biological clock makes daytime sleep far less effective than night-time sleep. That the regulations have still not changed is a classic demonstration of the prevalence of machine-centered thinking in this industry.

Another area of regulatory misguidance is the commandment—and federal regulation—"Thou shalt not nap in the cockpit." It sounds great in theory; who wants the pilot asleep at the controls? However, especially on night flights, many aircrew members take brief naps and the scientific data from studies conducted by NASA show that those who do nap while at high altitude are actually more alert during the critical final approach and landing than those who do not. The problem is that such behavior is forbidden by regulation and therefore is an illegal behavior. A captain who runs a flight deck by the book has a fatigued and sometimes dysfunctional crew who may inadvertently nod off even a few feet off the ground on the final approach. A captain who condones illegal napping behavior in the cockpit may have a better-functioning crew at the time of landing the plane.

Uncontrolled napping, however, is just as much a problem as banning napping altogether. Because napping is illegal, no procedures exist to make sure that it is done in a safe and carefully controlled manner. The subject cannot be regulated if it is banned.

Fortunately, the situation is changing. The convincing data from the NASA study are now leading after several years toward a rule change to allow controlled napping.

OPPORTUNITIES FOR AIRLINE INDUSTRY LEADERSHIP

THE AIRLINES, who are the operators of the aircraft, the employers of the aircrew, and the purchasers of the planes, will play a pivotal role in this revolution in thinking about human alertness and performance. It is an extremely competitive industry, and aviation safety is high ("The sky doesn't rain airplanes" was how one senior airline executive put it to me recently), so there will be natural curbs on any changes that cost money to implement.

However, much can be done that does not cost much money.

Having a sophisticated understanding of the day and night rhythms of the bodies of the aircrew is as important to airline management as having a good technical understanding of the cockpit instrumentation or the air traffic control procedures for shepherding planes through the sky.

Where to start? An obvious place is in the improvement of scheduling algorithms that generate the trips that pilots bid on each month. Right now there is relatively minimal input from human alertness/sleep-wake cycle considerations. To be sure, the FAA flight-time/duty-time regulations are heavily represented in the scheduling algorithm, but these, as we discussed, have little actual bearing on the realities of human fatigue.

More is needed than regulatory guidance. In fact, there is real concern about the value, and even the danger, of certain regulations, which if followed to the letter may excessively fatigue the unwary.

Currently, the computer programs of the scheduling department orchestrate each month a massive juggling act or optimization exercise. The routes the airline plans to fly (based on market research, travel volume, and gates owned at the various airports) are fed into the computer, as are the number and types of available planes and the aircrew positions needed to fly those planes. Taking into consideration the optimal use of crew, equipment, and regulatory limits, the computer spits out a bid-pack. Typically, the aircrew of the airline then bid for certain trip sequences in seniority order. Some trips are relatively easy, some are grueling. Some unnecessarily fatigue the pilots. A computer program is only as good as what you put into it. Until we input information relating to human alertness and circadian rhythms, dangerously fatiguing trips will be spewed out with high-tech efficiency.

Right now some review of trips takes place based on the complaints of the aircrew who are unfortunate enough to fly them. But no real equal weighting is given to these human considerations, even though they are just as encodable in the software routines.

The inertia of the mind-set of the airline managers is part of the problem, but it is also true that the economics of the aviation industry create strong resistance to anything that may

increase costs. Flying a pilot who is within legal flight-time limits, but who is liable to be a vegetable in the cockpit seat
because of fatigue, may save significant sums when it avoids the
necessity of having a replacement pilot available at a remote destination. But at what cost? Most of the time the airlines can get
away with it, but it is a game of Russian roulette, with innocent
lives and a $500 million accident at stake.

Not all of the resistance to change lies with the airlines.
Pilots also have their own agenda. Although many pilots are
greatly concerned about aircrew fatigue and want to see enlightened and better-informed scheduling routines in place, an equal
number like the benefits of the old routine. Look at the seniority
bidding, where the most experienced pilots may pick the most
grueling trip sequences because they maximize time off. Pilots
may choose to fly long, mind-numbing international trips with
many flip-flops of the sleep-wake schedule because they can get
a two-week uninterrupted break each month in which they can
pursue another business, investment, or hobby. This is not good
for flight safety, and is not good for the health of the pilot, but
the pilot may consider it worthwhile.

The remarkable commutes that aircrew often make to their
place of duty must also be factored in. Many think nothing of
flying the length or the breadth of the country in order to live
where they choose, and travel for them is typically free. Also,
many pilots love flying so much that they fly their own private
planes on their days off. Many also fly with the Air National
Guard to earn extra pay and to keep alive crucial hands-on
flying skills that may not be practiced in highly automated aircraft, especially by second officers who do not get to otherwise
handle the controls. Thus, to look at the overall impact of
fatigue, one must include all these extra flying hours in assessing the total challenge to alert and safe flying.

There are also major interindividual differences in the ability
to function well on a demanding flight- and duty-time schedule.
Some people can manage one type of trip sequence much better
than another, whereas for others the reverse may be true. Currently, no formalized efforts are made to match trips to individual differences, although it is possible to do, with great benefits
to all concerned.

TRAINING TO MANAGE FATIGUE

PILOTS CAN do much to improve their alertness on the job and to gain good-quality sleep in off-duty hours. They can also do much to make matters *worse,* without realizing it.

These issues come to the fore in night flying and international flying, in which sleep-wake schedules are disrupted and hours normally spent in bed are spent in the cockpit. But they are also true for daytime flying, especially if the hours are long and many time zones are crossed.

Aircrew must know as much about the causes of human error, and the important role of fatigue and loss of alertness, as they know about mechanical or procedural failures that can cause accidents. With human error accounting for two-thirds of aviation accidents, it is remarkable that it is given such short shift.

Federal Express has been the first to play a leading role in addressing the human fatigue problem. The company encouraged research on its aircrew by both the Institute of Circadian Physiology and by a team from NASA on pilot alertness and sleep patterns, and it commissioned Circadian Technologies to help develop the first aircrew training program on pilot fatigue.

The Circadian Technologies team worked with the Federal Express training team to develop a module for their regular recurrent training program. Sandwiched between cockpit procedures and technical updates, it has become a popular part of their training program. All Federal Express pilots now rotate through the module every year, and it is planned to expand the topic and add further content to it.

It is clearly time for all airlines to offer such programs, to ensure that their crews are as sophisticated in managing their own bodies as the machines they fly. Not only is it an essential contribution to aviation safety, but it is a tremendous contributor to improved labor-management relations. As long as airline management ignores this issue, they show that they do not realize or identify with the very real challenges aircrews face in trying to live an around-the-clock lifestyle. Few steps taken by management can be as cost-effective in improving morale and safety as providing such a training program and addressing the problem of human fatigue head-on.

Chapter 7

Medical Care Whenever You Need It

MY OWN RUDE awakening to the realities of the twenty-four hour society began after I had graduated from Guy's Hospital Medical School and began my internship in a major Toronto hospital. Until that time I had occasionally stayed up all night studying or partying. But largely this was an optional experience, and if I nodded off for a few minutes or was a social zombie, nobody cared.

When I arrived at the hospital on July 1 to begin my internship, the realities of my life for the next year rapidly became clear. I was to cover every other night, but also every day as well. That meant coming on duty at 7:00 A.M., staying on throughout that day and the following night *and* the day thereafter—thirty-six hours at a stretch. Then, after twelve hours to collapse and attend to the rest of life, I was back again on duty for another thirty-six-hour stretch.

In the old days doctors usually got an opportunity to sleep in the hospital for at least a few hours while on duty. Nothing much, other than an occasional emergency, happened at night. But that is no longer true. Fewer patients die quietly in the night these days, but at the same time hospital resident doctors get fewer hours of sleep. Modern high-tech medicine, with its intensive care units, electronic monitors, and alarms attached to every vital body function, makes sure that every deviation summons medical attention.

So there I found myself, expected not only to stay awake throughout day and night, but to be fully functional, too. I was expected to make life-or-death decisions, and do it with my sleep bank balance on empty at three o'clock in the morning. I still remember with a shudder the orders I am told I gave over the phone in the middle of the night, which I could not recall and could not understand the following morning. I remember standing hour after hour in the operating room in the wee hours of night, holding a retractor during emergency surgery, and having the surgeon yell at me because I had nodded off on my feet. And I was by no means the worst case. A California resident fell asleep while sewing up a woman's uterus—and toppled into the patient. In another California case, a sleepy resident forgot to order a diabetic patient's nightly insulin shot, prescribing instead another medication, which sent the man into a coma.

It is said that the price of training each young doctor is the loss of a patient's life, a life that could have been saved. From my own experience, I believe that this has to be true, such is the effect of fatigue compounded by inexperience.

It comes down to simple failures of human attention, energy, morale, and motivation. When you are dead on your feet and your body craves sleep, you tend to take shortcuts, to not make that final check on an IV, to skip a step and not double-check how much you have drawn into the syringe. Monday morning quarterbacking after a life has been lost says that the cause was human error, implying irresponsibility and guilt, when more accurately it was the health care system's failure to acknowledge organizationally the limited capacity of the human machine to respond.

A SYSTEM THAT CREATES FATIGUE

SURGICAL RESIDENTS in one Boston teaching hospital have been pulling stints of more than sixty consecutive hours on duty, in order to get some semblance of a weekend off. Work schedules of 130 hours a week—out of the only 168 hours available in a week if one never slept—have been repeatedly documented.

As a senior hospital resident described it:

I entered medical school with lofty aspirations of selfless dedication to the sick, a life of joyful altruism. Like many of my colleagues, I saw these worthy aspirations transformed into cynicism under the pressure of sleepless call nights in a pediatric internship. My love for children became a loathing. As an intern, I worked over 135 hours a week, taking short call (until ten o'clock) one night, long call (forty hours) the next night, and eight hours off the next night. I feel that sleep deprivation seriously impaired my judgment, giving me the experience of a sleep-walking nightmare.

Nurses tell the same story of chronic fatigue. As one nurse said:

Our hospital always seems chronically understaffed, so that they keep asking me to work double shifts. I can sustain my energy and my competent care for an eight-hour shift as an intensive care nurse, but I slide down hill, make silly errors, nod off when I should be paying attention, when I have to work sixteen hours straight. The holdover after a 3:00 to 11:00 P.M. evening shift for another eight hours through the night is the worst; at the end of that stretch I can barely think.

If you are going to get sick in the United States, don't do it at 4:00 A.M. in July—certainly not if you need urgent care in the emergency room of our teaching hospitals. July is when everyone switches to the next rung up the promotion ladder—when young medical students become interns and treat patients for the first time, and when interns become residents. At four o'clock on those early July morning, fatigue and incompetence dance hand in hand, which is not to say that one is free of such risk in other months of the year or hours of the day.

Why has the medical profession—the profession that ought to be most knowledgeable about human physiology and its limitations—adopted work practices that lead to breakdown of the human machine?

DANGERS FOR THE
HEALTH CARE PROVIDERS

IT IS NOT only the patients who suffer the effects of the inhuman schedules that pervade the medical world. In an environment where fatal diseases can be transmitted by an event as simple as a pinprick, small fatigue-induced errors of inattention can have fatal consequences. The most publicized case was of the resident physician Dr. Virginia Prego, who at the end of a long, exhausting night shift pricked her finger with a needle containing HIV-infected blood. She is now HIV-positive, and she is not the only one.

In other cases sleep deprivation leads to serious injury or death because of a highway accident on the way home. Consider the example of the young resident at the Tulane University Medical Center who was so fatigued by sleep deprivation that she dozed off at the wheel while returning home from a forty-hour shift. She is now quadriplegic.

Severe emotional stress is also brought on in health care professionals because of their fatiguing schedules. New physicians, after the initial exhilaration of their first job as a doctor has worn off, become somewhat depressed during their first-year residency because of the incredible combination of sleep deprivation and life-and-death job responsibility they must endure. What is less well known are the remarkable number of suicides and serious depressive episodes in these talented young people.

For other physicians and nurses the stresses translate into drug and alcohol dependency. Because of their access to pharmaceuticals, some find it all too easy to try to self-medicate their fatigue and stress. A senior and very well respected doctor on the staff of the Massachusetts General Hospital told me he had used cocaine to keep himself going through long stints on duty. The result can be a severely impaired person, whose ability to serve his patients is endangered.

The nursing and other hospital staff who work rotating shifts over many years also suffer from the problems of shift maladaptation syndrome, as discussed in Chapter 5. They have sleep disorders, gastrointestinal illness, and mood disorders as part of the general pattern of ill health that shift work can induce in those who are susceptible. Their bodies pay the price of their devotion to their job.

THE RATIONALE FOR MISERY

WHAT POSSIBLE justification could be used to support the 36-hour days and 130-hour weeks of the resident physician, or the two back-to-back eight-hour shifts of the hospital nurse? Are these a strange oversight of an overworked hospital administration or a conscious plan to torment?

The system that provides the young trainee doctor with an immersion experience into around-the-clock medical care has many vocal advocates in the medical profession. It has been the custom and practice for more than one hundred years to have young physicians work around the clock so that they can see and learn about the natural history of illnesses and the effects of medical intervention. The idea is that they can keep track of individual patients, provide a continuity of care, and at the same time allow the senior physician to sleep at home mostly undisturbed.

The trouble is that the modern high-tech hospital is a very different place from the hospital of fifty years ago. The bleeping monitors demand around-the-clock attention, the most complex surgical procedures now extend more than twenty-four hours straight. The expectation has been raised that equally competent medical help will be available whatever the time of day or night. The focused attention on one case evolving its "natural history" is too rarely possible; much more probable is coverage of multiple patients one has never seen before, and on which one has only the briefest details of medical history.

A perfectly valid concept has become bastardized by the need to provide physician coverage at the lowest costs and with the fewest staff. How else could a hospital pay such low salaries and have a physician on call? Just think about the numbers. A junior physician may receive only $25,000 per year and work up to 130 hours per week for fifty weeks per year—6,500 hours per year in the extreme case. This works out to $4 per hour—less than a person flipping hamburgers at McDonald's earns.

In part because of the low cost, physicians in training spend too much time on tasks that do not need a medical degree or even sometimes a nursing diploma. Inserting IVs, doing blood cultures, transporting patients, or scheduling tests consume too many of the interns' hours when they should be getting recuperative sleep.

Proponents still argue that sleep-depriving night duty is a valuable learning experience, that they did it early in their careers and suffered no permanent damage. But repeated thirty- or forty-hour stints of nonstop duty, for a year? It seems more like a rite of passage, a hazing by fatigue.

The ultimate responsibility of the physician is to the patient—to ensure the optimum in medical care whenever it is needed. But the unfortunate practice of staffing the front line with the fatigued and inexperienced does not meet a reasonable standard of responsibility.

ATTEMPTS AT REGULATION

THE BIGGEST single stimulus for change has come from the Libby Zion case in New York. When this eighteen-year-old woman died in 1984 after a night of inattentive care by fatigued and inexperienced residents in one of New York's major teaching hospitals, her father, an attorney and a former reporter for the *New York Times,* campaigned vigorously for change. Based on the findings of a Manhattan grand jury, which concluded that Libby Zion had received "woefully inadequate" care and had suffered repeated mistakes by first-year interns and second-year residents who had had little sleep, Sidney Zion managed to stimulate Dr. David Axelrod, New York State's health commissioner, into action.

One of the most remarkable and forceful health commissioners that New York—or any other state, for that matter—has ever had, Dr. Axelrod looked into the issue of resident physician fatigue and vigorously fought for limits of on-duty hours. The grand jury had recommended that regulations be promulgated to limit consecutive working hours for interns and junior residents in teaching hospitals, and Axelrod set out to put that into practice.

Against strong opposition by cost-conscious hospitals and by some senior physicians who thought it all bunk, Axelrod helped implement in 1988–1989 the following regulations:

- The emergency room shift for doctors should not exceed twelve hours.
- The *scheduled* workweek in critical care specialities should not exceed eighty hours.
- No *scheduled* shift should exceed twenty-four hours.
- Breaks of eight hours off should be allowed between duty shifts.
- One twenty-four-hour period off should be provided each week.
- Moonlighting should be banned.*

These limits are not particularly remarkable, and not reassuring for a patient being seen by a young physician who has been on duty for twenty-three hours straight. But the regulations created a furor: "Interference with the medical profession," "impossible cost burden," the cries went up.

The costs of implementing these new regulations for the New York Health Department were determined to be $65 million per year for replacing the lost cheap hours of hospital physician labor on nonclinical tasks, and $80 million per year for the clinical tasks that required extra medical staff. Another $81 million per year was estimated for increased supervision and bureaucracy. A scheme was devised to distribute these costs among Medicaid, Blue Cross, and the commercial insurance companies.

So far so good. But this was only for the State of New York, and lawsuits have since been filed and won by Blue Cross and

---

*If their work hours are reduced, many residents are tempted to earn extra money by working extra hours elsewhere, in their effort to pay back the enormous costs of medical education.

the commercial carriers so that the system is $50 million to $60 million short per year—largely because the federal medicare program does not pay its fair share. Furthermore, although resident hours have clearly dropped and progress is being made, regulations are consistently being evaded in some hospitals by using the loophole provided by the reference only to "scheduled" work hours in the regulations.

Because of the strong resistance of the health care system, even such modest changes in work hours have been slow to be implemented elsewhere. To head off threatened legislation, Massachusetts teaching hospitals developed their own set of guidelines that were not dissimilar to the New York regulations. But major national professional groups, including the Accreditation Council for Graduate Medical Education—which accredits teaching hospitals—have failed to set standards limiting hours. General surgeons and thoracic surgeons in particular have vigorously campaigned against allowing junior hospital physicians even twenty-four hours off once a week or placing any limits on the number of consecutive hours on duty.

Elsewhere in the world, similar battles are being fought against even the most minor restrictions. In the United Kingdom it took a successful lawsuit by a doctor against his employer, the Bloomsbury Health Authority, to prompt the minister of health to support an agreement to reduce doctors' hours on duty. The English Court of Appeal refused to dismiss the doctor's claim that the more than eighty-eight hours per week he was forced to work were unbearable and that they deprived him of sleep to the point of posing a danger to his own health and that of his patients. As a result a ministerial task force has now set a limit of eighty-three hours per week in the short term and seventy-two hours per week in the long term.

Interestingly, New Zealand has led the way in these reforms. In 1985 the New Zealand Resident Medical Officers Association successfully obtained a seventy-two-hour-per-week limit on doctors' hours. The results have been striking. On average, resident doctors' hours on duty have fallen to fifty-four hours per week—19 percent below the regulatory limit—and, contrary to predictions, no serious scheduling problems have arisen.

New Zealand's doctors are more awake, alert, and enthusiastic about their jobs, and patients have noticed improved care. Productivity and the maintenance of patient records have improved, especially on the night shift. To cap it all, the medical training experience has improved because doctors at their own volition may stay at the hospital to follow a particularly interesting case and watch the natural history of the disease and the effects of medical intervention, free of routine chores or responsibility for care.

RETOOLING OF THE MEDICAL MIND-SET

ISN'T IT STRANGE that the medical profession, with all its intimate knowledge of human vulnerability and the fleeting quality of life itself, should be so resistant to changes that enhance the effectiveness and health of human body and mind? A doctor flying to a medical meeting would not think it tolerable if the pilot of her airliner had been on duty for thirty-six hours straight. She would be most concerned on the highway if she knew that the driver of the truck approaching her was similarly fatigued. Yet because of the very real and immediate problems of her own profession, she suffers from an inertia of the mind-set so engulfing that she will campaign against the small step of reducing thirty-six-hour on-duty stints to no more than twenty-four hours!

What possible reason could there be for not seeing the obvious absurdity of such a position? One reason may be the overriding concern everyone involved in health care has about the containment of costs. The price tab has been climbing dramatically in recent years, so how can anyone consider changes that will increase those costs still further? A good question—but the wrong question, for in medicine we have been seeing the same phenomenon as we see in other parts of the twenty-four-hour society. The fascination with and the catering to technology is obvious. CAT scanners, MRI, PET, and a host of other technologies have transformed modern medicine but created an enormous capital investment challenge—not the least because U.S. physicians practice defensive medicine in a litigious society.

Even small hospitals purchase expensive technology, and with it comes an army of technicians to operate and service it.

No wonder there are no funds to spare for reducing human fatigue by a wiser deployment of personnel so that alertness and performance are cherished and enhanced. Somehow the human contribution is devalued even in this profession, which is the quintessential case of humanity serving humanity. Just as in the aviation industry, the gains made by technology are sorely compromised because of human fatigue and inattentiveness among the frontline troops.

A radical reevaluation of all factors that influence the alertness, performance, and health of health care personnel is needed. A good start would be a major revision of the job description of nurses, hospital resident physicians, and support personnel so that tasks are performed by a well-designed team that is in optimal condition of alertness at all hours of day and night. The night shift should not be treated as a "grin and bear it" exercise. It needs to be planned with a full understanding of the physiology of human alertness.

The agenda for change should include the following:

1. a personnel deployment plan that matches training level with the services to be provided and determines how the skill level of members of the team can be enhanced to support the patient care demands on the team;

2. advanced training of experienced nurses and technicians so that they can assume a greater level of responsibility for patient care within the new personnel deployment plan;

3. development of carefully designed shift systems that allow proper adjustment of the individual's sleep-wake cycles to the duty roster (replacing the primitive and unnecessarily exhausting schedules in use today);

4. comprehensive selection and preshiftwork training programs for all around-the-clock personnel to ensure that they are fully aware of how to adapt their own bodies most effectively to the demands of shift work;

5. design of work spaces, such as nurses' stations, to aid sustained alertness even in the middle of the night; and

6. development of supervisory systems that maintain the effectiveness of the team.

Is this a radical prescription? Not in the least. Our consulting group has been implementing these changes in hundreds of industrial plants throughout North America for nearly ten years with great success. However, one of the slowest industries to adopt such strategies has been health care.

Chapter 8

The Power
of Our
Society

I WAS IN the office of the vice president of
nuclear operations at a major U.S. utility company, talking about the risks of human fatigue and inattention
on the night shift in nuclear power plants, when news bulletins
started to arrive from the Soviet Union. A nuclear explosion of
enormous magnitude had occurred at 1:35 A.M. on April 26,
1986, at a nuclear power plant near the town of Chernobyl.
Never had a consultant's warning been underlined with such
force and such timing.

The first scraps of information out of the Soviet Union were
sketchy at best. But thanks to the developing *glasnost* policies, a
much more detailed account finally emerged. By far the best-
researched and most comprehensive source is the former Soviet
scientist Zhores Medvedevs's *The Legacy of Chernobyl*.

Human fatigue clearly played a big part in the Chernobyl
disaster. The explosion occurred while a special test was being
conducted under the supervision of an exhausted team of electrical engineers who had been at the plant for at least thirteen
hours, and probably longer because of a ten-hour delay in obtain-
ing permission to start. When they did begin at 11:10 P.M., a
series of catastrophically poor judgments were made. In a cas-
cade of errors, a control room operator on the night shift entered
the wrong data so that the rods were lowered too far into the
reactor, causing a loss of power. The test should have been
aborted then and there. Instead, the tired engineers tried to
hurry the test along by ordering a series of steps to boost power,

including deactivating some safety mechanisms. Then, to get still more power, they compounded their errors by ordering the rods to be pulled up—ultimately resulting in a runaway reaction and the steam and hydrogen explosions that followed.

The hot debris of the Chernobyl reactor deposited twenty million curies of radionuclides over five thousand square kilometers of surrounding land, rendering it uninhabitable. It also contaminated, to a lesser degree, large areas of northern Europe. In areas such as Wales and Cumbria heavy rains deposited significant radiation, contaminating food, water and livestock. With cleanup costs estimated as high as $300 billion, with 300 people dead so far and 1.5 million others contaminated with radiation, this was the worst-case scenario come to life. Underlying it were two classic errors that fatigued people make: entering the wrong data, and making poor judgments in the rush to complete a task.

Lowering the rods in a nuclear reactor is a slow and tedious process. The specific rod to be moved and the level to which to move it must be read from long charts of numbers. A fatigued person can transpose digits, inadvertently skip lines, enter the wrong data. The classic time such errors occur is in the wee hours of the morning when the human brain is most likely to be in a haze. Indeed, I visited one nuclear power plant where such errors had been made by operators three times in the past two months—resulting in a plant shutdown each time. In each case the event had occurred between 3:00 and 5:00 A.M.

The poor judgment to continue a task despite its imprudence shows obvious similarities with the fateful decision to launch the space shuttle *Challenger,* to be discussed in Chapter 10. In each case a fatigued group of people, under pressure to get a major task accomplished because of delays in the schedule, abandoned caution to the winds.

DEVELOPING AWARENESS OF THE RISK OF HUMAN FATIGUE

LESS THAN a year after the Chernobyl disaster, the U.S. Nuclear Regulatory Commission (NRC) sent a shock wave through the U.S. nuclear power industry by closing Peco's Peach Bottom nuclear power plant in western Pennsylvania. This action, on March 31, 1987, was taken because the control room operators had been discovered sleeping in their big comfortable chairs during the night shift. Never before had a nuclear power plant been closed because of human inattention even though no nuclear accident had occurred. As the late Republican senator John Heinz from Pennsylvania put it, the Peach Bottom closure was "a wake-up call for the nuclear industry."

The Peach Bottom plant was not allowed to reopen for another two years, with enormous costs and consequences for Peco and its management team. Peach Bottom reactors 2 and 3 had been supplying enough power to light the whole of Philadelphia, and $14 million per month of replacement power had to be bought from other utilities. The NRC fined Peco $1.25 million, the largest fine ever assessed against a nuclear power plant, and also fined the sleeping control room operators $500 to $1,000 each. All members of the management chain responsible for Peco's nuclear power operations were relieved of their jobs, including the company president and the chairman and chief executive officer. Even after the Peach Bottom plant had reopened following an extensive retraining of the crews, in which we assisted, a $150 million shareholders' suit against the chairman and CEO and the president of Peco remained. Two insurance companies finally agreed to pay $34 million to settle the shareholders' derivative suit.

The education of the nuclear power industry on the risks of human errors of inattention had actually started a few years before the Chernobyl and Peach Bottom incidents. In 1979 the control room operators at the Three Mile Island nuclear power plant near Harrisburg, Pennsylvania, were working the night shift at 4:00 A.M. when the multiple warning lights on the panels suddenly came on, indicating a serious plant malfunction. In their attempts to fix the problem, they overlooked a simple cause—a .007-inch valve had stuck—and because they missed it, their attempts to correct the problem resulted in a near meltdown of the plant. An alert day crew, arriving at 7:00 A.M. after a refreshing night of sleep, instantly saw what the real problem was, but by then serious damage had been done, both to the plant and to the reputation for safety in the nuclear industry.

THE CHALLENGE OF MAINTAINING OPERATOR ALERTNESS

WHAT IS IT about operating a nuclear power plant that makes the night shift such a challenge? Why is it hard to keep awake and attentive even when there is so much at stake? Once again, the enormous engineering effort that has been made to ensure nuclear safety through automation and control systems has itself created a work environment so bland and boring that it saps the alertness from the human machine.

The remarkably intensive training and simulator practicing that nuclear power operators receive—up to 20 percent of their time is spent in the classroom—creates a breed of racehorses with rarely any race to run. When all systems are A-OK, when the turbines are humming on and on, when everything is nominal, night after night, hour after hour, the idea that something could go wrong seems remote and theoretical.

The day shift is usually a hive of activity. Maintenance and engineering efforts to check out systems and conduct repairs require the operators to be on their toes. Circuits must be isolated and tagged so people can work on them, reports must be made to management, new employees must be trained. The control room may have as many as twenty-five people milling in and out, talking, working at computer terminals, replacing spools of paper in recorders, fixing things, holding conferences, speaking on the telephone or on walkie-talkees. The day shift can be exhausting to work, especially if the plant is being brought up or down from full power.

But on the night shift, the crowd drifts away, and the conversation between the two or three remaining people, while animated at first, becomes steadily more monosyllabic as the night wears on. Even if the operators are diligently trying to stay awake, they can drift into a tuned-out haze. I know because I have sat there beside them desperately trying to stay awake so I could accurately record on our observation charts what was going on, but still finding that the clock on the wall had a mysterious tendency to jump forward ten to fifteen minutes in an instant of time!

Just as in the chemical industry, the engineers and managers who supervise and regulate nuclear power plants sometimes create more problems than they solve in their extraordinary efforts to achieve the highest levels of safety. For example, they decide to reduce the level of distraction to improve the attentiveness of control room operators. How? They ban all reading materials from the control room except for the engineering manuals. But have you tried to read an engineering manual when your brain is tired? Nothing will put you to sleep more quickly. The NRC commissioners also tried to ban radios from the control room. Again, this is a commendable but misguided attempt. Think of what keeps you awake when you are driving on a lonely stretch of freeway late at night—your car radio, of course. And in the car, at least the scenery is moving!

Another misguided directive in the attempt to increase the level of "professionalism" restricts conversation to only that directly concerned with the running of the plant and the performance of operators' duties. When everything is running smoothly and quietly, there is little to talk about. I would much rather have two operators keeping their minds ready and alert by talking about baseball or anything else that interests them and keeps the brain alive.

The distinction missed in all these backfiring attempts to regulate is the difference between what is appropriate when there is an emergency or a critical procedure to perform, and what is appropriate when all is going monotonously well. In the emergency or during the critical procedure, of course, no radio should be on, only relevant reading materials consulted, and conversation directed to the vital task at hand. But the other 95 or 99 percent of the time, such a restrictive policy will serve only to disengage the clutch of the human brain so that it drifts off into neutral.

Do not blame the nuclear engineers and managers. They are merely part of a culture locked in an outdated paradigm—the same culture that has doctors working thirty-six-hour shifts and bans catnaps by airline pilots; a culture that fails to understand the essence of human-centered design.

PIONEERS IN RETHINKING HUMAN PERFORMANCE

BUT THERE ARE positive signs, such as research recently sponsored by the Nuclear Regulatory Commission. The Institute for Circadian Physiology has just completed a large NRC-sponsored study of the safety of twelve-hour versus eight-hour shifts. The industry research and development organization, the Electric Power Research Institute (EPRI), also commissioned Circadian Technologies to prepare an operating manual on "Control-Room Operator Alertness and Performance in Nuclear Power Plants"—which has become their "best-selling" manual. Furthermore, in our experience a greater proportion of nuclear power plants than any other type of around-the-clock facility have adopted strategies to improve shift scheduling and human alertness.

The willingness of many utility managers to admit that problems exist, coupled with a concerted effort by them to tackle these problems head-on, has been most welcome. This has occurred despite the dampening effects on openness that antinuclear activism has caused in the United States; for although such activism may foster public consciousness of possible dangers, it tends at the same time to make it harder to fix any problems that may exist.

The approaches we have used to improve human alertness and attentiveness in nuclear power plants as well as nonnuclear "fossil fuel" plants have been broad in scope because of the systemic nature of the problems. There is still much to do:

• Selection and Training

It costs approximately $280,000 to select and train an employee for four years to become a nuclear reactor operator, and to continue the training for six years to maintain the employee's license. Such an investment makes it extremely prudent to screen potential employees for susceptibility to shift maladaptation syndrome to avoid costly dropouts. Employees should also receive training on how to manage a shiftwork lifestyle, learning to minimize fatigue and sustain alertness on the job.

The nuclear power industry has been a leader in simulator training, where realistic copies of the control room are used to play out for teams of operators various types of problems or accidents and test their ability to deal with them. The training is exhaustive and impressive—except for one thing: it is not conducted in real time. The operators, who know when a problem is about to be given them, are like runners waiting with rapt attention for the starter's gun. A more realistic and instructive approach would be to conduct the training at night and create a boring simulation of a plant at full power; hours later, when attention is lapsing, throw the trainees subtle problems without warning to determine their response. This should be coupled with training exercises to enhance alertness and demonstrate their effectiveness.

• Managing for Human Alertness

The safe operation of a nuclear power plant requires a comprehensive rethinking of policies, procedures, and supervision that is fully aware of human design specs and the sources of human error. About 75 percent of the approximately three thousand reported incidents each year in U.S. nuclear power plants have human error as their principal cause, and errors of fatigue and inattention underlie many such events even when this cause is unrecognized. Management structures need to be set up to maintain human performance, to place human performance concerns on a par with technology and equipment concerns, to train management teams, to revise organizational structures. The job responsibilities of line managers and shift supervisors should be revised to include a key responsibility for sustaining alertness of their crews, especially on night shift.

• Scheduling and Overtime

The number of hours on duty and how they are scheduled is a key issue to be addressed. Such issues are subject to much pressure from the social and family concerns of the shiftworking operators and from the operational needs of the facility. Short-sighted cost savings can be achieved by understaffing the facility or not creating enough shift crews and then filling in the gaps with overtime. The plant saves money in training and benefit costs, and the operators take home a much bigger paycheck, but the real costs are hidden—the increased fatigue and risk of human errors of inattention. Effective shift scheduling strategies are now available, and there is no excuse not to adopt them.

• Man-Machine Interfaces and Work Environment

The considerable human factors efforts that have been undertaken on workplace design since Three Mile Island unfortunately have been lacking in understanding of the physiology of keeping people alert. Many surfaces are highly reflective sources of glare, encouraging dimming of the lights at night, temperatures are often too warm, and automation leads to human "zombiefication," a problem urgently needing a solution. New initiatives are under way to bring state-of-the art computer console driven interfaces and distributed control systems into the nuclear power plant environment to replace much of the 1960s and 1970s generation control equipment that predominates today. However, we need to watch carefully whose "state of the art" is introduced. It would be a fatal flaw to build the control room of the 1990s without a clear understanding of the human alertness technology discussed in Part III of this book.

• Regulatory and Oversight Overhaul

Two major forces are at work in the external governance of nuclear power plants. The Nuclear Regulatory Commission is responsible for the federal government's regulation and inspection; and the Institute for Nuclear Power Operations (INPO), an Atlanta-based industry-supported technical consortium, is responsible for inspecting and accrediting nuclear power plants. Both have a strong influence on policies and procedures.

As part of the game plan to reduce human errors of inattention, the policies and regulations promulgated by these organizations need to be brought up to speed. We discussed earlier how shortsighted policies that interfere with plants' attempts to ensure human alertness need to be revised. Restrictions on shift scheduling that prevent effective solutions need overhaul. Ideally, the NRC and INPO should become innovators along with EPRI in promoting the mind-set change of the nuclear power industry.

ALERTNESS IN NON-NUCLEAR OPERATIONS

NUCLEAR POWER plants represent only a small fraction of the around-the-clock operations of the utility industry. Much larger in number are oil and coal fired (fossil-fuel) and the hydro-electric, geothermal, and solar power stations, not to mention the many distribution and transmission systems and their control centers. Also vital are the linemen who clamber up poles and on whom we rely to return the power to our homes when power lines are interrupted by a storm-fallen tree.

The hours on duty and the risks of fatigue and injury are actually greater in these operations than in nuclear power plants, yet they do not have the catastrophe news value of the nuclear energy accident. Linemen, for example, work extraordinary hours under dangerous conditions; one lineman recently was fatally electrocuted after over eighteen consecutive hours on duty in responding to a storm-caused outage. In one plant we found operators working fourteen days in a row with twelve-hour shifts, burning themselves out in the process. We have seen, as we consult to non-nuclear facilities, other enormous problems of staffing and overtime that must be addressed.

Chapter 9

Keep On Trucking

THE UNSUSPECTING drivers came to a halt just before Junction 6 on the northbound lane of the M42 motorway near Birmingham. It was November 6, 1990. Roadwork on the exit sliproad had caused a back-up of traffic, and cars were queuing to get off the motorway. Frustration about the delay was surely on the drivers' minds, but none would have realized that they had only seconds to live. For just behind them a fatigued driver was at the wheel of an articulated lorry carrying 20 tons of steel bars.

The lorry driver had been working long hours, and had had less than five hours' sleep for the previous several nights. The monotony of the motorway caused a dense fog of fatigue to descend on his brain so he was functioning on "autopilot". Despite 700 meters of clear uninterrupted vision between him and the last vehicle in the queue, and despite the flashing warning signs of the cars ahead, his foot never left the accelerator. He continued to drive at 65 mph into the rear of the line of traffic, crumpling and burning cars in his wake and leaving six people dead and a dozen more seriously injured.

A VIRTUAL carbon copy accident six months earlier seriously injured Gloria Estefan, the lead singer of the Latin pop group, Miami Sound Machine. She was travelling in her bus en route to a concert in Syracuse, New York, when an accident ahead of them brought their bus and other vehicles on the highway to a stop.

But coming up behind her was a truck driver operating on too few hours of rest. After a long run from Toronto to New Jersey the day before, he did not get into his sleeper berth until after 1:00 A.M. He had started again early that morning and by the early afternoon siesta time he was so drowsy that, despite 500 yards of clear visibility and his CB blaring accident warnings from other truck drivers, he ploughed into the back of Gloria Estefan's bus. She went to hospital with two broken vertebrae and her international concert tour was cancelled. She could not perform for a year, and a court awarded her $8.5 million in damages.

TRUCK DRIVERS are not the only people who doze at the wheel, and famous entertainers are by no means the only victims. Increasing numbers of people must drive during the hours when the human brain is most fallible. The expanding twenty-four-hour society pulls more and more people into manufacturing and service jobs that require them to commute to and from work late at night and very early in the morning. To ease congestion, trucks are now banned from some cities during the day and hence must operate at night. And now that restaurants, bowling alleys, supermarkets, and launderettes are open around the clock, there are more reasons to go out in the middle of the night.

The rapid growth of just-in-time delivery, a machine-centered concept if there ever was one, adds to this around-the-clock pace. Companies can greatly reduce the amount of inventory or supplies that must be warehoused by having delivered only what they need precisely when they need it. Such an obviation of warehouse space saves serious money for manufacturing plants, but it adds to the burden on human fatigue. Now the truck driver is expected to arrive within an hour of any requested time of day or night to keep the assembly lines rolling.

FALLING ASLEEP BEHIND THE WHEEL

NOT ONLY the rare, idiosyncratic individual succumbs. Surveys of British motorists show at least a third have experienced episodes of nodding off behind the wheel. With people who work around the clock, the numbers reach 80 percent or greater. Sometimes when I speak to audiences in the bus or trucking industry, I ask how many have nodded off behind the wheel. I remember one five-hundred-person audience at the American Movers Conference where virtually everyone raised a hand.

Two studies, of the Oklahoma Turnpike Authority and from the California Division of Highways, suggest that as many as 50 percent of the *fatal* accidents on *freeways* are caused by drivers falling sleep or briefly nodding off. Such accidents constitute only 15 to 20 percent of all freeway accidents, so it is the most serious accidents that drowsiness causes.

When a person falls asleep behind the wheel they do not brake or swerve to avoid an obstacle as an alert person would. They just hit the bridge abutment or other vehicle full-speed, head-on, with no self-protective action.

In fact, investigators can pinpoint sleep as the cause of the accident when there is a smooth trajectory off the road or across the median strip with no skid marks visible. If the road takes a bend the vehicle goes straight, following its unintelligent path to its destruction.

The problem of loss of driver alertness and of accidents caused by microsleeps is enormous, yet little attention has been paid to it. Highway deaths caused by alcohol have, rightly, received much attention, but falling asleep behind the wheel is a problem at least as great. Indeed, in a recent major study of heavy truck accidents, more than half were found to be related to fatigue, significantly more than the number of accidents in which alcohol or drug abuse was to blame. Since medium and heavy truck accidents in the United States total $9 billion in damages per year, this little-talked-about problem of truck driver fatigue is costing us approximately $5 billion a year. Imagine the figure if we included the cost of all the *automobile* accidents caused by driver fatigue.

Why the lack of attention to fatigue-caused accidents? No blood or breath test can be performed afterward to show that fatigue was to blame. Furthermore, the drunk driver is still drunk after the accident, but the driver who fell asleep may

Figure 9.1

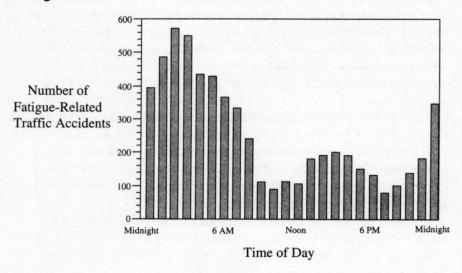

Number of Fatigue-Related Traffic Accidents

Time of Day

show no sign of drowsiness. To the contrary, unless dead or severely injured, the driver is likely now wide awake and alert! But more important, we tend to see the accident caused by a sleepy driver as without fault, an unfortunate occurrence, whereas drinking too much is considered a socially irresponsible act. Yet there can be as much damage, injury, and death from a fatigued individual as from a drunken one.

When the EEG brain waves and EOG eye movements of free-way drivers are measured, characteristic drowsiness patterns can be observed. Bursts of alpha, theta, and even delta sleep may intrude into the waking EEG. Of particular interest are microsleeps, brief five- to ten-second episodes where the brain is effectively asleep while the person is still driving with his or her eyes wide open. My colleague Dr. Claudio Stampi at our institute has been recording microsleep episodes of truck drivers, and even commercial bus drivers, out on the highway. Microsleep episodes occur in bus drivers with a full load of passengers behind them, truck drivers with nuclear weapons on board in the confidential unmarked convoys of the Department of Energy's nuclear courier service, and mothers transporting their precious offspring.

The drowsiness comes in waves. Although you may be reasonably alert one moment, you can sink into a drowsy spell quite rapidly, especially, as we discussed earlier, during the postlunch hours and in the middle of the night—the times when fatigue-caused accidents most frequently occur (Figure 9.1). After fighting the urge to sleep for a few miles, respite usually arrives, encouraging the traveler with a deadline to just keep on trucking. But this driver's account is a chilling cautionary tale:

I was driving my truck through the night trying to make a deadline when these bouts of drowsiness started to come over me. As I always do, I grit my teeth and tried hard to pay attention. But then I started to feel a strangely cold sensation under my leg. I reached down and found to my surprise I was wearing no pants and had cold hard metal beneath me. I couldn't figure what was going on so I looked down and saw my clothes in a bag next to me. Suddenly I came to the realization that I was in a hospital emergency room and that I had been in an accident.

The very toughest time to stay awake on an overnight trip is at dawn just as the sun rises. As the first rays of light peek over the horizon, the tired brain rebels. Single-vehicle truck accidents—a useful measure of driver inattention—occur twenty to thirty times more frequently at this time than during midmorning hours. Wise truckers stop and sleep, even if just briefly, at this time, but the demanding business schedules of our society too often do not allow such prudence.

REGULATIONS THAT FORCE YOU TO CHEAT

OVER FIFTY YEARS ago, concerns for public safety led to the adoption of laws that prescribe the on-duty time, the driving time and the rest time of commercial vehicle drivers. In the United Kingdom a lorry driver may drive for no more than nine hours a day (ten hours are allowed two days a week) and then he must have eleven hours off per twenty-four hours. In the United States driving hours are limited to ten, to be followed by eight hours off before driving again. Sounds like a reasonable first step, doesn't it? But consider what this means in practice: Say you drive for ten hours, take eight hours off, and then drive for another ten hours and take another eight hours off. You find yourself on an inhuman schedule—your work-rest pattern cycles every eighteen hours instead of every twenty-four. You are forced into a topsy-turvy life of flip-flopping between day and night driving and day and night sleep.

The human sleep-wake apparatus works best when kept on a regular twenty-four-hour schedule. The human brain works least well on irregular patterns of sleep and wake. On schedules such as that described above, drivers find themselves alert when they are mandated to be asleep, and driving legally when their brains are fogged in by fatigue. Jim Johnston, the entertaining and outspoken president of the Owner-Operators Independent Truck Driver Association, tells of his early experience as a trucker, when he and a young colleague were each assigned a load from the Midwest to New York City. His colleague, who followed the Hours of Service regulations to the letter, arrived late and dead tired. Johnston, who disregarded the regulations and slept when he was tired and drove when he wasn't, arrived fresh and alert and on schedule.

To circumvent those regulations, most truckers keep two sets of logs—one for inspection purposes, strictly by the book, and another to keep track of the hours actually worked. This universal practice makes a mockery of the regulations, which not only deserve to be mocked but should be overhauled.

Of course, we do need regulations. Irresponsible drivers will otherwise drive eighteen hours straight, for they are making money when the wheels are rolling, when loads are delivered in the shortest possible time. Such behavior creates a serious public safety hazard from which the driving public must be protected. But to overregulate, and to wink and nod at the subverting of regulations, doesn't help. Once breaking of the law is accepted, there may as well be no law.

The Canadians introduced regulations that permit up to thirteen hours of driving but require a longer rest period—at least encouraging the driver to maintain a near twenty-four-hour day-night schedule. U.S. studies are being conducted, still in their early stages, to explore the problem of truck driver fatigue. However, they focus mostly on the fixed-route, terminal-to-terminal (less-than-truckload, "LTL") truck drivers, and not on bus drivers or the much larger number of truck drivers who drive on irregular schedules.

Regulations do not and never can govern the automotive motorist, a major source of fatigue-related accidents. Overall, regulation is a poor solution, and alternatives need to be developed.

DRIVER EDUCATION

DRIVERS ARE their own supervisors. No one watches over them, telling them when their performance has deteriorated and they should drive more safely. Professional drivers therefore need special training to learn the physiology of alertness, the steps they can take to sustain alertness on the job, and how to decide when to rest, how to use caffeine wisely, and how to schedule eating and sleeping when out on the road.

The nine switches of alertness can be used to fashion a special training program for the professional truck driver. It is an effective way to reach the decision makers in this industry—the drivers themselves.

Take, for example, the steps to follow when you are extremely drowsy. Most people know how to flick on the temperature switch of alertness by turning up the air conditioner or rolling down the window. They know how to turn on the sound switch by tuning the radio to a channel that will keep them awake, or how to flick on the stimulation switch by changing the speed at which they are driving. But what if all this is not working, your head is bobbing, yet you cannot call it quits for the day? Many people stop their vehicle, get out and jog around for a few minutes, and then climb back in and drive away. Although this may improve the situation for a while, once you are out on the highway the sleepiness returns. Indeed, one study looked at people driving round and round a large test track who, when they felt drowsy, either kept plugging along or stopped and got out, stretched, and walked around their car before proceeding. Those that continued to drive encountered the next severe drowsiness bout, on average, twenty-seven minutes later; those who stopped and walked around encountered their next bout, on average, within twenty-three minutes—that is, no improvement was seen.

A cup of coffee can help to reset alertness for a longer period of time. Caffeine resets the chemical stimulation switch—provided the person is not too fatigued and not overdosed with caffeine already. Another effective alertness boost is to pull off the road at a rest stop and take a *brief* nap—ten or fifteen minutes is ideal and can refresh you without leaving you groggy. *Avoid a longer nap,* which can leave you worse off than before.

TECHNOLOGICAL HORSE SENSE

IN THE DAYS of the horse and cart, one was at least protected by the modicum of common sense shown by the horse. The motor vehicle in its current configuration shows no such sense; without human intervention it will not hesitate to drive itself straight into a wall. Two approaches—sensing the environment and sensing the driver—can alleviate this problem.

The classic engineering approach is to devise systems that can detect approaching collisions and issue alarms to the driver. In aircraft, a similar concept was used in the design of the system that triggers a warning voice saying "Pull up, pull up" if the cockpit altimeter senses the plane is approaching the ground too fast. Although a considerable challenge, this approach is well liked by engineers because it involves physics and engineering. It makes much more sense, however, to diagnose the source of the problem—the driver behind the wheel.

Early efforts in this direction are being explored by a number of companies, including Nissan and Mazda, and are beginning to be seen in the vehicles of the 1990s. Today's systems measure human behavior—the patterns of steering wheel movements and pedal pushing. There are also devices, which don't work very well, for detecting head-bobbing. Tomorrow's systems will be more sophisticated and tap into the physiology of the individual.

Just as we saw with airline cockpits and oil-refinery control rooms, increasing driver comfort presents unforeseen dangers. In earlier times, before paved roads and iron railways, travel was an intrinsically uncomfortable process whether one was on horseback or bumping up and down in a carriage along potholed roads. Nowadays we make it much easier to fall asleep and hurt ourselves while traveling. Modern engineering has given us comfortable, shock-absorbing vehicles with cruise control that ride on smoothly paved freeways. We also travel at much more dangerous speeds. It is almost as if we have conspired to fail.

A truly human-centered automobile would be designed to sustain the performance of the command center—the driver's brain. Part III of this book will introduce the technology to detect deteriorated alertness and correct it. But before this technology can become incorporated, a major mind-set change in the thinking of automotive engineers must take place.

ALERTNESS ON THE IRON ROAD

FOUR HUNDRED AND SIXTY-EIGHT COMMUTERS boarded the 6:14 A.M. Poole to Waterloo and 906 boarded the 7:18 A.M. Basingstoke to Waterloo on Monday December 12, 1988. They were preoccupied with completing their crossword puzzles or planning for the day ahead. But thirty-five of them were never to make it past Clapham Junction and another 69 were to spend weeks or months in hospital.

At 8:10 A.M. the driver of the Poole train passed a green signal light and entered into a left-hand curve through a cutting just before Clapham Junction station. But as he rounded the bend he was confronted with the unthinkable; the Basingstoke train was stationary just ahead on his track. There was insufficient distance for him to stop. Despite applying full emergency brakes, he ploughed into the rear carriage of the Basingstoke train. Moments later, an empty train heading outbound from Waterloo to Haslemere collided with the wreckage spread over the tracks.

Why had the signalling system failed on one of the busiest railway tracks in England, where trains pass at speed every two minutes during rush hour? The answer ultimately lay in

British Rail's gross understaffing of the Signals and Tele-communications Department's technical team which was responsible for installing a new signalling system. As a result 28 percent of the signalling technicians were working seven days every week, and another 34 percent were working thirteen days out of every fourteen. Sometimes they worked thirteen or more hours a day with only a five-minute break. Through overtime, the technicians could more than double their take-home pay.

It is scarcely surprising that a disastrous error of human inattention would creep into such an overworked situation. The senior technician who incorrectly wired the signal which misled the train drivers had taken only one day off in the previous thirteen weeks. All it took to cause the disaster was his oversight of an uninsulated wire in the relay room in Clapham Junction "A" signal box. This wire was to short-circuit the signal. With no adequate staff or procedures to cross-check a fatigued employee, the accident was destined to occur.

IF ERRORS of inattention can occur in wiring a signalling box, imagine the challenge to the train driver on long overnight runs. There is much less stimulation to keep you awake than driving a car or truck. The responsibility for steering is taken away—the iron rails controlled by the switchmen do that for you. There are no lights, no other traffic, and, most important, little in the way of an immediate sense of danger. You can stop paying attention for ten seconds, even a few minutes, unless you are on a highly trafficked line—something you can obviously never do on the highway.

Yet there is very real danger. Other trains are on the track, sometimes stopped ahead of you, protected only by signal lights that you are expected to see. The hours of work are even worse than those driving a truck. Particularly problematic is the practice of keeping railroad engineers on a shift roster waiting at the terminal to be called in sequence for the next available train. With apologies to William Shakespeare, to sleep or not to sleep, that is the question. They may never get a train on that shift, and if they sleep they will confuse their sleep-wake pattern off duty at home. If they don't sleep, and several trains unexpectedly arrive at once, they may find themselves called out at the end of their shift on a run that may exceed twelve hours—and they will end up in a severely sleep-deprived state.

The classic time for railway accidents is in the early morning hours. Burlington Northern had two such accidents within nine days. In one, two railway workers were killed at 4:45 A.M. when both the engineer in the lead locomotive and the head brakeman fell asleep, and they missed several warning signals before plowing into the back of another train. In the other accident, five railway workers died when two freight trains collided head-on at 3:55 A.M.. In these accidents, twelve locomotives and sixty-one cars were damaged, for a total cost of more than $5 million.

Length of time awake can also be a key issue. In November 1990 four railway workers were killed in a similar accident involving an engineer who had been awake for twenty-six hours and forty-one minutes at the time of the accident. Not only freight trains are at risk. Amtrak has its share of such accidents, although little attention seems to be given to engineer drowsiness in these investigations; in contrast, mechanical failures or drug or alcohol abuse inappropriately receive much greater attention.

What actually happens when a train is rolling along through the night has been beautifully demonstrated by Tjorborn Akerstedt and his colleagues in Sweden. They wired up railroad engineers with EEG, EOG, and EKG electrodes on the run from Stockholm to Malmo. During the day runs, alertness was usually well maintained, but the EEG brain waves were diagnostic of a much different state at night. Alpha, theta, and delta waves intruded and the slow rolling eye movements of drowsiness were observed. Of eleven engineers studied, six were shown to be dozing while driving the train, and two missed warning signals because of microsleeps.

In one episode, a train engineer was recorded as he slipped into a microsleep and rode through a warning signal without applying his brakes. The brave researcher, riding on the train, watched the frightening evidence spew out from the recording machine, but in the interests of science did not intervene. A little while later, the engineer came out of his haze to find a red stoplight immediately ahead of him. He slammed on the brakes, with heart literally pounding as evidenced by a doubling of the rate on his electrocardiogram trace.

The rules of work hours need to be overhauled in the railroads as they do on the highway, and education on alertness management is just as essential. The technology being developed to track the status of the engineer is interesting. Many trains have installed systems that require engineers to press a button every so often to prove that they are alert. If they do not press it, a warning buzzer sounds, lights flash, and finally a klaxon sounds that would wake all but the dead or the dead drunk. If all else fails, the system stops the train—a luxury not available to highway drivers.

It's an aggravating system, which engineers have learned to bypass. Better solutions are on the way, as discussed in Part III.

DANGERS OF FATIGUE ELSEWHERE

WE HAVE discussed the problems that nonstop operations present for the aviation, health care, utility, and transportation industries. But these represent only a small sampling of the businesses and industries of every type imaginable now being challenged by the transition to a twenty-four-hour society. Similar chapters could be written about paper mills, oil refineries, semiconductor plants, and candy factories, or about retailing, banking, insurance claim processing, stockbroking, or print and electronic media. More still could be written about military operations by land, sea, and air, or security operations, police forces, fire, and other emergency services—each of which runs continuously around the clock. Each has its own special set of problems and opportunities.

The themes of this book reverberate through almost every area of human activity. For some industries, such as oil refining, the success in automating continuous processes has led to the creation of high-tech "sleepy hollows" in which the challenge is to keep operators awake and alert for the infrequent occasions when intelligent human intervention is required. For other operations, such as in heavy manufacturing, the varying tolerance of the human body of physical fatigue throughout the course of day and night may be the cause of wear and tear on the shiftworking employees.

Employees in maritime operations, for example, have notoriously demanding and irregular work patterns that disrupt their sleep and leave them dangerously fatigued. An assistant bosun who fell asleep on duty failed to close the bow doors before the *Herald of Free Enterprise* set sail from Zeebrugge harbor for Dover. One hundred and fifty passengers and thirty eight crewmembers lost their lives when the boat capsized. Employees in other occupations, such as oil-field technicians, are required to sustain performance over many consecutive hours in remote terrain. Even those who work strictly nine-to-five hours and sleep comfortably in their beds at night are not free of the side effects of human failure in the nonstop world. The foggy-brained mechanic who omits a step in checking your plane's engines, delaying your airport departure; the autoworker who zombies out as your car-to-be passes by on the assembly line, wasting your time and money for repairs because of that missing spring; the fatigued operator on the night shift miskeying your insurance claim—all contribute to a loss in your personal productivity and sometimes your safety.

We are all victims of the accumulated sum of fatigue-induced human failure. Even those at the top do not escape, as we will see in the next chapter.

Chapter **10**

Decision-Maker Fatigue

THE YOUNG EYES of the world were on the NASA space shuttle *Challenger* on January 28, 1986. It was the first mission with a schoolteacher on board, and she planned to conduct lessons from space. With the launch already delayed, there was an urgency about this mission, and thousands of NASA engineers and contractors were working overtime to get the spaceship into orbit. Finally, the space shuttle roared off from the launchpad, only to explode seventy seconds later against the bright blue sky, in front of a stunned nation and packed classrooms of horrified children.

There had been a perfect opportunity to avert the disaster late the evening before. A key prelaunch teleconference was held between the senior engineers and managers of NASA's Marshall Space Flight Center, Kennedy Space Center, and of Morton Thiokol, Inc., the makers of the booster rocket that would explode. They failed to pay sufficient attention to the risk of launching in colder temperatures than the rocket was designed to withstand.

Unfortunately, the key managers who participated in the fateful decision to launch were fatigued by exceptionally long hours on the job. The day before the launch, many of them arose early, several before 3:00 A.M. after only two or three hours of sleep. By the time the key go-ahead decision was made late on the evening before the launch, several had been up for more than twenty hours. These highly responsible but tired people did not make the appropriately deliberate decision in the face of their personal urgency to "hit the sack" to get ready for a busy day ahead.

As a result, the U.S. manned space program ground to a halt. The confidence that NASA had instilled in the American belief that all things are technically possible was dealt a stunning blow. But it was not the technology per se that failed; it was the human decision of when to deploy it that had gone awry.

WE HAVE focused so far on the frontline troops of the twenty-four-hour society—the shift workers, airline pilots, truckers, doctors and nurses—but what about the decision makers at the top? The presidents, prime ministers, corporate chieftains, and generals, and their executive decision-making staffs and senior management teams. How are they affected by the demands of a nonstop world?

Do the power and resources at their disposal protect them from the vicissitudes of the twenty-four-hour society? Are they able to determine their own schedules, sleep at sensible times, decide when Air Force One, or the corporate jet, will fly? The answer may be yes in theory, but too often the opposite is the case. Instead of providing a way for senior executives to take personal advantage, the technological power of our world allows those in leadership positions to further abuse their bodies and compromise their ability to be fresh and sharp.

INFORMATION OVERLOAD

IN PREVIOUS eras the inefficiencies of communication allowed time for adequate thought and contemplation. Even in war, operations ceased at night in medieval times, except for the very occasional surprise attack. But nowadays there is no such excuse—the information flows in ceaselessly in real time in so much quantity and quality from so many places around the world that there is never a pause for breath in the crisis situation room.

The temptation is to be always tuned in to CNN, always be plugged into the flow of information about fast-changing events.

When everything seems so important and so vivid, it seems irresponsible to take time out to sleep. But that is precisely what a responsible leader should do. When leaders make decisions in the fog of fatigue, the enhanced power that technology has provided amplifies the magnitude of any errors made. The impact of an army sent to invade, a nuclear weapon launched, a hostile takeover instigated, or even an international diplomacy initiative botched, reverberates through thousands or even millions of lives for many years thereafter.

One of the most remarkable demonstrations of how the communications revolution has changed our world was seen during the Persian Gulf War. Ordinary American citizens sitting at home in their armchairs watching live reports on CNN of the direction of incoming Iraqi Scud missiles were on the phone relaying the information as it came in to their relatives and colleagues in Saudi Arabia and Bahrain, warning them to take cover.

Think about that for a moment. You and I today have far more information and far greater powers of communication at our fingertips than world leaders like Roosevelt or Churchill had at any time during World War II. Live TV feeds, instant telecommunication, satellites, and computer links provide an amazingly instantaneous sense of what is happening around the world. If the average citizen can be privy to so much information, just think what global leaders have at their disposal. But the extrapolation of the relationship between an individual's power and the volume of information he or she can assimilate is nonlinear. Leaders put their pants on one leg at a time like the rest of us, and their brains absorb facts and process images at not very different rates. In fact, it turned out that the decision makers around the world during the Gulf War were watching CNN much of the time, just like the rest of us.

But they had a responsibility to act; we were only armchair critics and voyeurs. They had to cope with this barrage of information; we could switch off when we got tired, and we could go to work fatigued from staying up to watch the compelling TV images. They were fatigued from the same lost sleep but had to make coherent sense of it all and act on it the next day.

FOGGY-HEADED IN FOGGY BOTTOM

IN SUCH international crises, tremendous dependence is placed on clearheaded decision making by these national leaders and their close advisers. Yet as a crisis brews with information coming in minute by minute around the clock, the strong temptation for decision makers is cut back on sleep and stay on their feet until they can function no more.

As the days and nights pass, accumulating sleep deprivation and fatigue renders these key people less and less able to make the best decisions on which the world relies. They may become irritable, irrational, and tempted to make shortcut decisions rather than carefully evaluate all the potential outcomes. Indeed, their need to stay involved, participate in every decision, and be briefed on every development undermines their very effectiveness in crisis management. How appropriate it is, then, that the State Department in Washington, D.C., is located in an area sometimes referred to as "Foggy Bottom"!

We must learn to rethink our attitudes about expectations for leaders. The press is all over a president who is not woken up from sleep with news of any incident the press deems important. President Reagan was lambasted for not being awakened on every such occasion, blamed for being out of touch. The implications are clear from the media and those who read them every day; we expect our leaders to be sleep deprived or else they will be deemed slothful.

Short of an incoming nuclear missile attack on American territory, little justifies waking a president from sleep. When an important incident occurs, it is usually much more important for the senior decision makers during their regular workdays to be alert, clearheaded, and deliberate in their actions. Very few crises require instantaneous responses, however instantaneously the crisis information is being received.

A lesson learned by human factors experts after the Three Mile Island nuclear power plant accident should be taken to heart. The serious problems arose because fatigued people acted in haste. There would have been no disaster, no near meltdown, if the operators had done nothing and taken their hands off the controls.

Not that a government should do nothing quickly when a crisis arises, or that a major oil company should sit idly by when one of its tankers founders on the rocks. Any major twenty-four-hour operation with a critical mission must be staffed by senior and experienced crisis managers on a coverage schedule around the clock. It is essential to have people available at any given hour of day and night who can call the shots and activate the first-level response, and whose work and sleep schedules ensure they are in a fit state at that hour. Then the strategic management of the crisis can be overseen by fully functional leaders.

When staffing around the clock, think carefully about the individuals you deploy. Some of us regularly get up out of choice at 4:00 A.M.; give us a 5:00 A.M. crisis, and we will be at our best. Others regularly stay up until 1:00 or 2:00 A.M. and are their most productive and effective at that hour. But put the late night type in the situation room at 5:00 A.M. or the early morning person there at midnight and you will get a less-than-stellar performance.

A person's ability to handle a tricky problem is highly dependent on his or her level of fatigue or alertness. The first functions to go are the creative "lateral" thinking. Problems that may seem overwhelming when one is tired trigger half a dozen possible solutions when one is awake and alert.

Wisdom says "I'll sleep on it" whenever a major decision is to be made. It is critical for the leader not to be browbeaten by anxious subordinates, or an anxious constituency, into making a decision he or she will later regret.

In the immediacy of getting a space shuttle launched, or a negotiation completed, it is too easy to push ahead regardless. But in hindsight, when things have gone awry, or the unthinkable disaster has happened, the challenges posed by a delay seem minuscule in comparison. Faxes may be pouring into the office, phones and videolinks may be besieging the decision maker. But to take charge means that you must also be responsibly aware of your physiology and show respect for the age-old rhythms of the human brain. It is not only obvious crises that can result in catastrophe due to decision-maker fatigue. Critical errors can occur when it is business as usual, but with too few hours in the day to do what one has to do. Vital data were overlooked by the managers at NASA and Morton Thiokol in the urgency of getting a decision made so everyone could get some rest. Indeed, it is at such times, when a crisis is not apparent, that decision-maker fatigue can have its most disastrous effects.

HUMAN LIMITATIONS
IN GLOBAL DIPLOMACY

GONE ARE the days when business was geographically localized or when international diplomacy was conducted at a stately pace. The technological links that draw the world ever more closely into a global village have created a sense of urgency just because they have made instant communication possible. But it is not the rapid transmission of electronic images and transduced sound that cause the worst cases of decision-maker fatigue. Rather, these occur when we transport ourselves physically to another time zone. Most people can struggle through a single translocation across time zones, but the international negotiator may go to an incredible extreme.

Witness the travels of the former U.S. secretary of state Alexander M. Haig, Jr., in his attempts to mediate the Falkland Islands crisis in 1982. Shuttling among Argentina, the United States, and Great Britain, he accumulated thirty thousand intercontinental air miles in a couple of weeks. In one eight-day period he could have adjusted his watch twenty-two times to a different time zone. After one eighteen-hour trip from Argentina, he immediately plunged into eleven hours of talks with British officials. Although Haig had a specially equipped Boeing 707 with a sleeping bunk at his disposal, this did not mitigate the effects of the extraordinary disruption of his sleep-wake pattern. Why is it that we let our senior diplomats and decision makers conduct important missions in a state compromised by fatigue?

One of the well-known early cases of jet-lag-botched diplomacy was that of another U.S. secretary of state, John Foster Dulles. In the 1950s he flew to Egypt to negotiate the Aswan Dam treaty. Although many factors contributed to the diplomatic breakdown, Dulles—fatigued and irritable from jet lag—clearly mishandled the sensitive negotiations. The project was lost to the Soviet Union, ushering in a decade of strong Soviet influence in that country.

And if you think it difficult for a world leader or diplomat to adjust, think about their support staffs. They do not have special lounges and sleeping bunks; they have to make do with the usual stresses of aviation travel.

Several years ago a young man came into my office and flashed a badge at me announcing he was from the U.S. Secret Service. I quickly wracked my brains to figure out why I or any of my colleagues or students might be thought a threat to the nation! But he had come to seek help with an extraordinarily bad case of jet lag precipitated by his job of protecting U.S. presidents and secretaries of state on their international missions. With total disrespect for the human biological clock, the U.S. Secret Service had hatched a duty roster that placed each agent on assigned shifts around the clock—shifts that ran always by local time wherever an agent was, creating—on top of the normal jet lag—an incredibly random work pattern. Agents would be on duty and then have their shift expanded by eight hours just because they happened to cross eight time zones. The agent who came to my office, and his colleagues, were totally burned out by this experience but were receiving no sympathy from their immediate supervisors.

As has been too frequently proven, the reactions and judgments of the U.S. Secret Service are critical to protecting the lives of American leaders. But their effectiveness will be compromised as long as the human bodies of the Secret Service agents are treated with less respect than one would treat a machine.

MAINTAINING DECISION-MAKER EFFECTIVENESS

THE NEXT sections of this book detail a philosophy and a variety of strategies for maintaining peak human performance, most involving a sensible awareness of human limits and designing systems and procedures to play up human strengths rather than weaken them. If true crises develop, where sustained attention is needed, a variety of techniques and technologies are available. Short multiple napping techniques can reduce total sleep need to as little as two to three hours per twenty-four for weeks on end without deteriorating performance; special bright artificial lights can suppress sleepiness even in the middle of the night; and performance can be enhanced by prophylactic napping. Scheduling of sleep and design of the sleeping environment can improve the quality of recovery sleep. And choosing the correct timing of waking from sleep can significantly affect grogginess (sleep inertia) on awakening.

This technical information needs to be disseminated to the decision makers in government. Myths of immunity from sleep deprivation need to be dispelled and effective policies instigated to ensure that our key decision makers are as free as possible from unnecessary fatigue, for in this ever more complicated world we need all the alertness and wisdom they can muster.

Chapter 11

Human Fatigue
and the
Law

MATTHEW THEURER, an eighteen-year-old
high school senior in Portland, Oregon, was
working after school at a McDonald's restaurant to earn some
spending money. He worked the evening shift until 11:30 P.M.
one Sunday and then went in after school on Monday to work
from 3:30 to 8:00 P.M. His employer needed help in cleaning the
deep-fat frying machines and offered Matthew an extra shift
that night from midnight to 8:20 A.M. The following morning,
after only seven hours of sleep in forty-eight hours, he climbed
into his car to drive home and fell asleep behind the wheel on
the way. His car collided with a station wagon, killing Matthew
and severely injuring the other driver. A case was brought by
that driver against McDonald's for negligence in allowing Matthew to drive home in a fatigued state. The jury awarded
$400,000 in damages; another $10 million suit is pending from
Matthew's family.

THE LITIGATION of accidents, injuries, and
errors caused by human fatigue is undergoing rapid change.
Long an archaic backwater of vague defense and weak claims,
unsupportable by hard evidence and deduced by dubious cri-
teria, fatigue was rarely a satisfactory approach to winning or
defending a legal suit. In a driving accident or a train wreck, it
was usually more satisfactory to search for a mechanical failure,
even if not directly the cause, because that was more tangible; or
to look for alcohol or drug abuse because that could be chemi-

138

cally detected even in a corpse. Without a similar chemical test for fatigue, it could not be proved with the same ease.

Today attorneys and the courts are becoming more aware of the impact of sleep deprivation on performance and the deterioration in attentiveness caused by fatigue. Furthermore, they are increasingly assigning liability to the employer, as in the McDonald's case. At the same time, the scientific research on alertness and sleep has formed the basis for a much more rigorous approach to analyzing such cases. Expert analysis and witness support can now be provided for legal cases, and an effective method for assessing what is loosely called fatigue has been developed. It is now possible to blast through the vague generalities and reach a level of clarity based on scientific research that can form part of an effective defense or offense.

SLEEPLESS NIGHTS FOR THE CHIEF EXECUTIVE

A SERIOUS problem for today's CEOs is the size of the judgments obtained in American courts of law. The *Exxon Valdez* spill made clear that there is virtually no upper limit to the damages assessed in environmental catastrophes. A multibillion-dollar liability assessed against a less deep pocketed oil company than Exxon could put it out of business. Or some hourly worker whom a CEO has never met, twenty-two steps lower on the corporate ladder, can doze off and create a disaster that will occupy the time and energy of that CEO for months. The dozing control room operators discovered by the NRC inspectors at the Peach Bottom nuclear power plant of Philadelphia Electric (Peco) set off a chain reaction that destroyed the careers of many managers and ultimately cost Peco's chairman and CEO, James Lee Everett III, and Peco's president, John H. Austin, Jr., their jobs.

Worse still for chief executives is the increasing trend to hold them personally liable for human errors by their employees. Everett and Austin became targets of a shareholder suit that assessed millions of dollars of liability for the shutdown of the Peach Bottom plant and the loss of revenues and profits. In another case, an Indian court recently requested the extradition of Warren Anderson, retired CEO of Union Carbide, to face homicide charges in the 1984 Bhopal disaster in which approximately four thousand people were killed from a methyl isocyanate chemical plant leak—even though Anderson was nowhere near the site and had no direct responsibility.

The CEOs and senior managers of many high-risk businesses are desperately trying to stem the tide. Because the stakes are so high, charges of fatigue-caused human error or accidents must be analyzed very carefully, using the best available tools for analysis and modeling using human alertness physiology. An in-depth look at these cases often reveals much more complexity than is at first apparent.

WHAT IS FATIGUE ANYWAY?

THE PROBLEM with the term *fatigue* is that the closer it is examined, the less it is understood. People think they know what they mean by the word *fatigue,* but how does one measure it in the context of assessing human performance impairment?

If you have taken a college-level physiology course, you know that muscle fatigue is a reasonably well defined concept. But we are thinking in terms of "brain fatigue"—the state in which a person cannot think straight or keeps dozing off in bouts of inattention, and as a result creates a safety hazard. Because it cannot be measured by objective means, this type of fatigue is of little use in a court of law or in the comparison of one situation with another.

As Lord Kelvin once said, "If you cannot measure something, you do not fully understand it." Without a precise scale of measurement, there is no way to determine whether a particular behavior, situation, or treatment alters the state by a predictable amount. Of course, if I tell you that a night without sleep leaves you with increased fatigue, you understand what I mean; but I cannot tell you in exact terms the minimum amount of sleep you need to avoid becoming fatigued.

That may be the type of precision needed to resolve a legal case. At issue may be the question of whether a person had obtained adequate rest, or whether an employer had procedures to ensure that an employee was fit for duty. Standards or benchmarks cannot be set without a quantitative scale by which to judge them.

ALERTNESS—A MEASURABLE STATE

DON'T LOSE all hope! There *is* a measurable state related to fatigue—and that is the level of "alertness." Fortunately, unlike fatigue, alertness can be measured relatively precisely as we saw in Chapter 4. Alertness—best understood as the state of activation of the brain—is a physiological state that renders your brain either a smoothly running machine or a barely cranking engine misfiring on all cylinders.

I can tell you precisely how much alertness is altered by having five versus seven hours of sleep the previous night. And I can tell you what impaired alertness means in terms of reduced performance or fitness for duty. I can measure, simulate, and testify to alertness. Loss of alertness is related lawfully to inattention, microsleeps, falling asleep on the job or behind the wheel, or tuning out into an automatic zombie state.

Now we have a level playing field and a chance to evaluate claims and counterclaims on a scientific basis. However, I still use the word *fatigue,* as well, because it is useful in connoting a generally understood concept. The only thing it lacks is a precise meaning.

WHEN TWO SHIPS DO NOT PASS IN THE NIGHT

EARLY IN the morning of October 30, 1988, the Carnival Cruise Lines ship *Festivale* slowly crept up the Anegado channel leading to San Juan Harbor. Passengers were starting to wake, and the early birds were getting ready to come on deck to watch the entry into port. The captain of the *Festivale,* one of Carnival Cruise Lines' most experienced skippers, had joined the crew on the bridge after a normal night of sleep to oversee the docking procedure that he had done more times than he could count in his forty-plus years at sea.

Dawn had not yet broken as the *Festivale* was met by the tugboat *Neill McAllister,* dispatched to assist in the maneuver to dock at Pier 6. But the seas were calm, and the pilot and *Festivale*'s captain decided they did not need the tug to nudge the vessel. Instead they directed the *Festivale* to proceed ahead under its own steam with the tugboat keeping pace behind. Less than an hour later, the *Neill McAllister* lay at the bottom of San Juan Harbor with three big gashes from the *Festivale*'s propellers in its side and its three crew members swimming for shore.

Claiming $1 million of damages for the lost tugboat, the insurers of the *Neill McAllister* filed suit against Carnival Cruise Lines. The tugboat captain claimed that his boat had been pulled into an unusual amount of wash from the *Festivale,* and had consequently swung into the path of the propellers.

Herb Brown, of Calvesbert and Brown, a leading marine law firm in San Juan, was retained by Carnival Cruise Lines to defend it. Early in his investigation, Brown noticed the large number of overtime hours the tugboat captain had been working and wondered whether "fatigue" might explain what really had happened.

When Brown called to seek my advice, my first step was to conduct an initial assessment of the facts of the case. Were there obvious factors that indicated significant impairment from loss of alertness on the part of the tugboat captain? It quickly became apparent that there were, in what turned out to be a textbook case.

TECHNIQUES FOR ASSESSING PROBABLE LEVELS OF IMPAIRMENT

MEASUREMENT of the alertness of responsible parties, using techniques such as EEG analysis and the multiple sleep latency test (MSLT), is obviously not useful *after* the incident. However, we can simulate the probable level of alertness because of the considerable amount of data that has been gathered using the MSLT. By constructing a computer model that simulates the ebb and flow of alertness, using MSLT equivalents as our measure, we can look at scenarios such as the *Neill McAllister* accident. When we did so, the considerable impairment of the tugboat captain became clear. In fact, the simulation showed that the accident occurred at the lowest point on the captain's sleep-deprivation depressed alertness curve (Figure 11.1).

Such analysis is much more than mathematical simulations. We also look for evidence for the existence of a set of factors that are indicative of the person's impairment. These include the following:

1. Hours of consecutive duty at time of accident

2. Hours of duty in preceding week

3. Irregularity of duty schedule

4. Time of day relative to biological clock

Figure 11.1
Simulation of Tug Operator Alertness

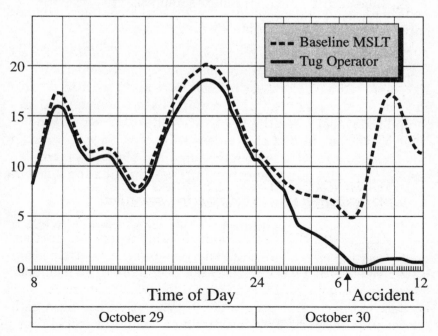

5. Predawn effect

6. Hours since last consolidated sleep

7. Duration of last consolidated sleep

8. Cumulative sleep deprivation over past week

9. Timing and duration of naps since last consolidated sleep

10. Fear/awareness of danger

11. Monotony/stimulation of job

12. Monotony/stimulation of social/physical environment

13. Prior physical and mental work load

14. Environmental temperature

15. Environmental lighting

16. Environmental sound

17. Environmental aroma

18. Ingested chemical stimulants/depressants

19. Physiological circadian/sleep-wake type

These factors alone are not enough to prove that loss of alertness caused the accident, however. A person may experience extremely reduced alertness but not have an "opportunity" to have an accident. For example, a ship's helmsman can nod off behind the wheel when the vessel is on the open sea with no other boats in sight and no accident will occur. So we also must look for mechanisms in which the impaired alertness is converted into accident-prone behavior. These include the following:

1. Loss of attention in a dangerous situation

2. Fixed focus on a minor problem, when there is risk of a major one

3. Failure to anticipate danger

4. Divided attention

5. Automatic behavior syndrome

6. Microsleeps in safety-critical situations

7. Failure to appreciate gravity of problem

8. Failure to observe warning signs

9. Impaired logical reasoning

10. Inappropriate corrective actions

By the time the accident occurred at 6:30 A.M. on October 30, 1988, the tugboat captain had been on duty almost continuously since 3:00 P.M. the previous day. He had worked the 3:00 P.M. to midnight shift on the *Neill McAllister* and then worked until 2:17 A.M. on another, more powerful twin-screw tugboat, the *Fritzy K.* He had then changed tugboats, and had a short rest of one or two hours before he was called out again at 5:00 A.M. to meet the incoming cruise ship *Festivale.*

It was dark when he met the *Festivale,* and as he trailed the ship into San Juan Harbor dawn was just beginning to light the horizon—the most deadly time of day for someone who has been up for most of the night. By the time the accident occurred, our mathematical simulation showed that the tugboat captain's sleep-deprivation depressed alertness level had dipped to its lowest point—much lower than that of a fully rested person.

The task of trailing immediately behind the stern of the *Festivale* required continuous attention, but it was an extremely monotonous job for the captain, all alone on the bridge of his tugboat. There was no steward to bring him coffee, the other crew members were down on the deck, no one could relieve him. He just had to hang in there and try to concentrate on keeping his tugboat steady in the stream of wash that would spin him around backward and into the stern of the cruise ship if his attention lagged for a moment.

And then it happened. The fog of a microsleep appears to have descended on his brain so that, for a moment, he lost his attention on the task at hand. Before he knew it, his boat was spinning around and being pulled toward the *Festivale*—and his reflexes were too slow to take the necessary corrective action. The tugboat swung around and came alongside the *Festivale.* The tugboat's bow, now pointing to the stern of the *Festivale,* became entangled with metal hooks on the side of the cruise ship. Obsessed with the potential ripping of the canvas on his tugboat's bow—a relatively minor problem—the captain failed to notice the large red letters just in front of him that warned of danger from the screws below.

Three loud thumps occurred below deck, and the captain still didn't realize what was happening until the tugboat started to take on water and sink. He and his crew jumped into the water and swam safely to shore, leaving their tugboat at the bottom of San Juan Harbor.

Not every case is so clear-cut, but this account shows how the role in a human error incident of factors that cause loss of alertness and reduced performance can be assessed.

WHEN SLEEPING ON THE JOB IS THE MOST EFFECTIVE DEFENSE

ON THE OTHER side of the fence is a case that initially looked hopeless but ultimately could be effectively defended. It illustrates that the single-minded emphasis on hours on duty—virtually the only factor regulated by the government—can lead to false conclusions.

This case took me to Artesia, New Mexico, and involved a truck driver who hauled water in a tanker truck back and forth for hours on end from a water tank to an oil-well rig that needed water urgently to keep it under control.

Early one evening in February 1988, the truck driver had made his eighth round trip of the day, which involved driving along 1.5 miles of side road, stopping to cross a divided two-lane highway, and then driving on another 3.5 miles to the oil well. This time he stopped at the divided highway, crossed the south-bound lanes, and then stopped in the median strip before starting to cross the northbound lane. Just as he did so a car came out of a dip in the highway and struck the side of his truck. The driver, a middle-aged woman, unfortunately suffered some injury to her back and ultimately sued the trucking company for damages, to which she seemed reasonably entitled given that she had the right of way.

The issue that involved me in the case was her attorney's decision to sue the company for triple damages because of the high
number of hours the truck driver had worked that week. They
alleged that, although the hours were legal under the regulations for local as opposed to long-distance trucking operations,
the company had caused the driver to be severely fatigued,
which in turn caused the accident. A well-known sleep expert,
called in by the plaintiffs, argued that the driver had worked too
many hours in the week before the accident and therefore must
have been severely sleep-deprived.

However, we carefully examined the facts and discovered several things that weighed heavily against loss of alertness as
being the cause of this accident. These included the following:

1. The accident happened at 7:00 P.M., the least likely time
for highway accidents to occur because of fatigue. Statistics
show that because of the properties of the sleep-wake system,
the probability of driver-fatigue accidents reaches the lowest
level at that time. One is least likely to succumb to a
microsleep or fall asleep at the wheel. However, some fatigue-
related accidents do still occur, so we cannot totally rule out
the possibility from this fact alone.

2. Although he had worked extraordinarily long hours in the
days before, the driver had had a full night's sleep the night
before the accident and had reported to work at his normal
time of day. Hence a significant amount of recovery of sleep
debt had occurred, reducing the impact of the long previous
hours of work and making a significant deposit into his sleep
bank.

3. Most important, our analysis of his job showed that the
driving schedule for each round trip provided him with multiple opportunities for short naps—once or twice per ninety-
minute round trip. It took thirty minutes for the tanker to fill
up and sometimes thirty to forty minutes waiting to unload at
the well site. His cab was equipped with a board between the
seats that enabled him to lie down. That the naps were short
was fortuitous, since it has been shown that multiple short
naps can replace the need for sleep and allow a person to get

by on many fewer hours of sleep per day. An added advantage is that you can wake up with minimum grogginess caused by sleep inertia.

Although this pattern of sleeping on the job was not devised with a sophisticated knowledge of alertness and sleep physiology, the company was fortunate it operated this way. Our analysis and testimony placed real doubt on whether the driver was truly fatigued at the time of the accident. Although limited damages were awarded, which gave the plaintiff a reasonable redress for her injuries, an inflated damage award was avoided by introducing an understanding of human alertness physiology into the case.

As the global twenty-four-hour society expands its reach, companies have to expect increasing litigation concerning human fatigue, not only as a contributor to accidents but also in worker compensation cases concerning ill health caused by shift work in some people. Increased precision in analyzing these cases and in defending against them will be a necessary part of corporate strategy in a world in which the incidence of fatigue in the workplace, on the highways, and in the air is on the rise. If left unchecked, this sea of litigation would bankrupt our economy and sap the vital energy and time of managers and executives the world over. It must therefore be contained both by reducing the risks and by handling the litigation with the utmost care and expertise.

III

THE
EMERGING
HUMAN
ALERTNESS
TECHNOLOGY

Chapter 12

Monitoring the Status of the Human Brain

FORTUNATELY, just when it is most needed, a critical human-centered technology is emerging to help address the problems of human frailty in our nonstop world. Human alertness technology offers the opportunity to control the state of activation of the human brain so that the goal of "Alertness When Required, Sleep When Desired" becomes a feasible target.

An intelligent, fully interactive human alertness technology first requires continuous quantitative information about the state of the human brain. Is the person, the user of the equipment, alert and firing on all circuits, or has the fog set in? Is the person paying attention to the task at hand, or is he attentive to some other distraction? Is his functioning impaired, by sleep deprivation, drugs, or alcohol, so that he is not fit to interact with the equipment?

Some of these characteristics vary from minute to minute, for the human brain is a very dynamic device. For utmost responsiveness, a human alertness monitor must be constantly scanning, to keep track of the fluctuations in human brain state as they occur. In contrast, to detect drug and alcohol impairment, the updating can be less often.

The optimal approach to tracking the alertness and attentiveness of the brain varies depending on whether the need is for fast responsive action or only occasional checks on state. Some methods are already developed and are entering the marketplace; others we know how to do but need investment of time and money to achieve.

FITNESS FOR DUTY

THE GROWING awareness of the enormous and potentially catastrophic consequences of human error in industry, particularly with the amplification effect of capital-intensive operations (see Chapter 1), has lent some urgency to the problem of checking whether employees are fit for duty when they arrive at work. Increasing awareness and concerns about drug and alcohol abuse, and recognition of the debilitating effects of sleep deprivation, have led employers to seek ways to determine whether an employee is likely to be dangerous to self and others—not to mention whether the employee is capable that day of efficiently performing the job.

Computer performance tests designed to test employee fitness for duty are now being developed and marketed. They are intended particularly for the so-called "safety-sensitive positions" (jobs where the consequences of human error are large). They are clearly preferable to urine testing. Employee acceptance is higher, and because the results can be obtained immediately, they are relevant to making decisions about that day's work.

This is the simplest form of human alertness/attentiveness monitoring. The decision is yes or no—the employee works or doesn't work. There is no attempt to modify or manipulate the level of performance capability that day, other than telling the employee to take a nap, or to go home and sleep off the hangover. Disciplinary action may often be triggered, with termination if the problem is repeated.

Such tests carry several limitations:

1. They determine the state only at the moment the measurement is made. A person may arrive for the night shift at 11:00 P.M. and pass the test but then be downright fatigued four hours later and not functioning well. Alertness states in particular may change fast, so a single snapshot of performance taken at the beginning of the shift does not protect the employer very well.

2. The employee can muster special effort to pass the test but then be impaired in the normal monotony of work. If one's job is at stake, paying attention for a few minutes will really pay off for the employee, at the employer's expense.

3. Employee acceptance may be a problem because the employer is playing a Big Brother role that employees may resent. If not well handled by management, the computer performance test could become a visible reminder that the employer does not think the employees are to be trusted.

4. There is always the danger that the computer test does not measure a type of performance that is relevant to the person's job.

Fitness-for-duty tests nevertheless can be a good first step in addressing a difficult problem. Certainly they are better than the alternative of letting a person do a safety-sensitive job in an already impaired state.

THE MEASUREMENT OF ATTENTIVENESS

WITHOUT attentiveness, the human brain is deprived of essential input and severely limited in the tasks it can perform. As with fitness for duty, it is of little value to check attentiveness only at the beginning of a work shift. The dynamic and changing state of attentiveness can be meaningfully measured only by continuous monitoring, or at least, monitoring at times when attentiveness really matters. During the process of landing an airplane or changing the positioning of the rods in a nuclear reactor, a person's state of attentiveness five minutes ago is yesterday's history if the individual is not attentive at the vital phase of a critical maneuver.

This makes measurement of attentiveness a significant challenge. Constantly alert and rapidly responsive must be the hallmarks of such a system. It must know when paying attention matters, and provide feedback that helps the sustenance of attentiveness without itself being a distracting element that could reduce attentiveness to the task at hand. We are in the process of developing such a device that is rapidly responsive, nondistracting, and information rich. Using special remote sensor technologies, we can obtain a continuous update on the attentiveness of the operator, without that person being aware.

For successful introduction into the around-the-clock workplace, the feedback method must be sensitive to the needs, perceptions, and attitudes of the monitored person. Constant watching by Big Brother is not a mental image that we want to foster. Rather, the system needs to be considered a helpful and unobtrusive aid that backstops the human operator. It cannot take away a person's job responsibility—instead it must be a supportive reminder and facilitator.

ALERTNESS MONITORING

ATTENTIVENESS without adequate alertness is not sufficient. The state of activation of the human brain—that is, alertness—should also be monitored in critical safety-sensitive jobs. Attentiveness by a foggy brain may result in impaired processing of essential information, dumb errors made, wrong buttons pushed.

But alertness monitoring cannot replace attentiveness monitoring, for a fully alert person who is distracted by some extraneous event or thought will not be effective on the job. The truck driver distracted by the sight of a pretty woman at a busy traffic intersection, your daughter doing her homework in front of the TV, and the sailor watching the dolphins play rather than the rocks ahead, are alert but not appropriately attentive.

Perhaps the most dramatic example of failure of attentiveness was the Eastern Airlines crash in the Florida Everglades. Crew members were so distracted by a faulty landing-gear position indicator that they ignored their altitude, and the plane crashed needlessly into the ground. Their alertness was stimulated by the potential danger posed by the perceived landing-gear problem, but their attentiveness to other more costly risks, tragically, was absent.

These two variables are not independent. With decreasing alertness comes increasing risk of failure to track multiple simultaneous risks in the environment. Attentiveness to more than one task at a time—the so-called "divided attention task"— becomes impaired. The fatigued driver with impaired alertness will fail to look in the rearview mirror when steering to avoid an obstacle in the roadway ahead. Most of the time he gets away with it, but not when he is being overtaken by another vehicle coming up behind him.

The development of alertness monitors is also feasible. With extreme and dangerous deterioration of alertness, sleepy events intrude into the EEG brain waves and EOG. K-complexes, those hiccups in the EEG, as well as theta and the slow delta waves, may occur in short bursts. Slow rolling eye movements may also be displayed, all warnings of alertness critically impaired.

The first generation of alertness monitors will detect such events and warn that a safety net is needed. Recent NASA studies of pilots in commercial aircraft show that they suffer such changes in their brain waves whenever they are sleep deprived or fatigued, even if they are on final approach to a landing.

Subsequent generations of such monitors will focus on giving as early a warning as possible. Too late a warning to take avoidance action is no warning at all, as demonstrated by some alertness warning devices now on the market. Placed behind your ear when driving, they detect, using mercury switches or the like, when your head nods forward, when muscle tone in the neck relaxes as you nod off. The problem is that you may have already gone off the road and hit that tree before such a device is activated.

More effective approaches to monitoring driver alertness may be the devices that the Japanese automobile manufacturers, such as Nissan, have been pioneering. These use the variability of steering-wheel movements as a performance measure that correlates with the alertness of the driver. It is well established that the fully alert and attentive driver has a pattern of constant, small driving-wheel movements; the sleepy driver has fewer such movements interspersed with larger movements as the driver jerks the drifting car back into its lane. Patterns of steering-wheel movements vary significantly between individuals, but by learning the characteristic behavior of each driver, the device can detect deteriorating deviations from each person's norm.

Better still, advance warning can be given by measurements of the state of activation of the brain. We expect to see great strides made in this area in the next few years.

CONSTRUCTING THE FEEDBACK LOOP

DETECTING danger is of little use if you do not take advantage of the information. Thus essential to the design and implementation of alertness and attentiveness monitoring systems are feedback systems that ensure good use is made of the information.

This is not as easy as it sounds. You might think that warning the human operator that she is dangerously sleepy might be enough, but merely telling a person she is sleepy may be an irritation, something she feels she already knows. The lighted coffee cup on the dashboard that the Nissan device uses as its warning signal may not be noticed or may be treated as a statement of the obvious.

Furthermore, if the sensitivity is set too high, so that the device keeps sending off warning klaxons at the first signs of reduced attentiveness, the device is likely to get "fixed." People are amazingly creative in finding ways to deactivate or circumvent devices that bug them.

The best approaches are more subtle. Unless the person has already fallen fast asleep, corrective actions that stimulate attentiveness and alertness are best. Increasing the work load, requesting the individual to perform extra but necessary tasks, is a way both to stimulate alertness and at the same time to diagnose performance impairment. It cannot be a meaningless exercise that the operator "sees through," but rather a set of tasks that must be done periodically anyway. The alertness/attentiveness monitors just select the times the operator has to do the task based on when operator stimulation and performance testing would be most useful.

Another approach is to use what we know about the factors that influence alertness. Bright lights and cool temperatures help switch on alert behavior. One can envisage systems being developed that automatically modulate lighting levels and ambient temperatures so that their fluctuations sustain the operator's alertness level. The changes might be so subtly triggered that the human will not be conscious of them.

Just like old Professor Cannon's recognition, that the autonomic systems of the body "removed the control of the essential functions of the body from the caprice of an ignorant will," so might these human alertness technologies take over the sustenance of alertness in the middle of the night because our conscious decision-making capabilities are not up to it. Sleepy people often do not have the mental or physical energy to take actions that would make them less sleepy. The recognition of an oncoming wave of sleep may just be a passing thought on the way to becoming fully submerged.

RETHINKING THE JOB DEFINITION

THE MORE power the engineer has to monitor and correct human attentiveness and alertness, the more realistic the engineer must be about human capabilities. I once was shocked by a conversation with an oil-refinery process engineer who defined the key requirement for the control room operator as the continuous monitoring, minute by minute throughout a twelve-hour shift, of the process control monitors. When was the last time you constantly monitored anything for longer than a few minutes?

We must be realistic about alertness and attentiveness on jobs, such as that of the oil-refinery operator, especially on the night shift. Any job, however responsible or critical, that involves monitoring a computer screen can require only brief, intermittent bouts of attentiveness, and a reasonably sustained, although fluctuating, alertness level.

Attentiveness/alertness feedback systems must be designed

accordingly, calling for fully attentive behavior only when necessary. Alertness has to be maintained at a state of readiness without expecting the full physiological flight-or-fight response at all times. This is essential to the effectiveness and acceptance of these systems by the employee at the man-machine interface.

USING MULTIPLE SENSES IS COMMON SENSE

ATTENTIVENESS should ideally focus on all relevant senses. The fully attentive operator will use sight, sound, smell, temperature, and vibration in tracking machine performance. We do it every day in operating our automobile while driving to work or to the store. The dashboard displays emphasize visual cues, but an abnormal vibration of the chassis, an unfamiliar smell, strange noises, and steam escaping from the hood are also well-recognized signs of trouble.

Less well appreciated is how much an oil-refinery operator, a machinist, or a paper mill technician also relies on a wide variety of senses in tracking equipment performance. Indeed, in this computerized age we have focused on visual cues because that's what computers do best—a classic case of machine-centered thinking.

We may be depriving operators of essential information by moving them away from their machine and placing them in a quiet room with a computer terminal. Certainly, the necessary information can appear on a computer screen, but a number or bar chart is far less compelling than a change in the vibration frequency in the floor beneath one's feet. An operator may be tuned out, engaged in a conversation, or not watching the computer screen but still instantly alerted by the vibration change.

It is hard for computer engineers to output information as vibrations, so they don't do it. But such information can make the difference between life and death. As the pilot reported in the story that opened Chapter 6, it was the change in vibration as the plane started to stall that aroused him and saved lives when the whole cockpit crew inadvertently fell asleep.

Computer engineers are only now attempting to make better

use of sounds as cues. Moving away from the simpleminded pre-computer gong and buzzer approach, sound synthesizers provide rich opportunities for auditory information, but they are as yet barely plumbed by man-machine interface designers.

Certainly unrecognized by engineers is the alertness-stimulating effect of having multiple, information-rich sensory cues. Remove a machine's vibrations, replace its reverberant whistles and gasps and rumblings with a computer console's constant white noise or a sixty-cycle hum, and eliminate the distinctive smells of the process, leaving the smell of the carpet—and you take away the factors that used to keep operators alert and responsive. Some key switches that stimulate the level of alertness have been flicked off.

Chapter 13

Boxing Up
the Sun

THE LIGHT SWITCH on the wall—the legacy of Thomas Edison—gives instant satisfaction. Darkness changed into light, night into day, with such ease and certainty that we almost cease to think about it. Only when a storm knocks out the electric supply and we are left in the dark, without radio or TV, toasters or hair curlers, computers or fax machines, hot water or electric stove, do we realize how dependent we have become. But soon the power is restored, and we think no more about this technological miracle.

It is, however, a "miracle" that we should examine more closely. Thomas Edison, and all the other inventors and engineers who made electric power and light a reality, gave humans the initial opportunity to work or play at any time of day or night. Economic pressures, market forces, and the human innovative drive took care of the rest. Factories opened around the clock to increase return on capital investment, services expanded into the night hours to meet market opportunities, the twenty-four-hour society was born.

But Edison's light bulb was not a human-centered technology. It serves human pragmatic goals but not human physiology. It provides the illusion that we have conquered the night, but we really have not. Electric light deviates in some essential ways from the natural light of the sun, whether or not filtered by cloud cover. The temperature, color spectrum, and, above all, intensity and pattern of onset and offset are different.

The light of the sun provides us, for example, outdoor intensi-

ties of at least 10,000 lux on a heavy, overcast day; 50,000 lux
when the clouds have cleared; and 100,000 lux or more on an
open beach or snow-covered ground where much light is
reflected from the surface. In contrast, the incandescents or
fluorescents of a brightly lit office at night provide only 500 lux,
a level that is subtwilight, a level that you would see in nature
only before the sun has crept over the horizon. Many industrial
workplaces provide even less light at night; 200 lux or less are
frequently seen.

Our control over night and over schedule that is provided by
the light bulb is only an illusion. However, human alertness technology can change this. Through examination of the sun's contributions to illumination of our environment, through the teasing
out of all biologically relevant features, it seeks to place the characteristics of the sun in a box. There is exciting potential here to
assist human adaptation, well-being, and performance in the
strange world our technology has inadvertently created.

BRIGHT LIGHTS, BIG CITY

IN THE PAST few years we have come to recognize how
bright artificial light must be to mimic the sun's effects. The
specs of the artificial lighting needed for the twenty-four-hour
society have come to be understood.

The first stunning revelation was that the human brain is
unlike that of most other animals in its response to light. Much
of our research uses mice, rats, and hamsters, not to mention
myriad other more obscure species. The biological clock of each
of these species is readily reset by low levels of light, and sup-
pression of the pineal gland's secretion of melatonin, a hallmark
of the night, requires no more than your common or garden light
bulb.

When similar levels of light, or even brighter levels of artificial light, were found not to suppress the human nighttime melatonin surge or control the timing of the human circadian
rhythm, it was first assumed that the human species had
evolved so as to escape the dictates of light and dark. Perhaps, it
was mused, humanity's discovery of fire was accompanied by a

revolution in the human genome, freeing human beings from
the dominating influence of the sun.

A few years ago, however, it was discovered that pineal melatonin could be suppressed by artificial light—but at light intensities five times greater than the brightest office light levels: the
same response as in other animal species, but with a greatly
reduced sensitivity to light. These light levels were not so bright
when compared to the range of those provided by the sun. These
light levels were equivalent to the first rays of day creeping over
the horizon. It seemed the human system was well protected
from moonlight, starlight, and even campfire light, but was still
responsive to the dictates of the sun.

These light levels, significantly greater than the lights of any
big city, kept our species in synch with the twenty-four-hourly
spin of the earth on its axis. They kept us bound to the schedule
of our ancestors, despite the invention of electric light and
despite our seeming mastery of the night. We had spread our
activity into the night, but our bodies had not followed suit.

Other discoveries rapidly followed. The depressive mood of
the shortened winter day in temperate climes, the Morgatid—
severe depression—of the arctic winter lightless day, could be
reversed provided adequately bright light (2,500 lux plus) were
provided in the early morning. The syndrome called seasonal
affective disorder, with its appropriate acronym, SAD, became
understood as a result of bright-light deprivation as day length
shortened in the fall and winter months.

How low the ambient-light levels really are for many people
in the modern world was also recognized. The elderly, especially,
see little light, for the electric bulb provides them the excuse not
to see the light of day. Night workers in winter months may
never see the sun, for it is shining only when they are resting
indoors.

How best to provide these levels of light? The first research
centers used huge banks of fluorescent bulbs, whole walls of
light—and lots of air conditioning to minimize the heating of the
room. But soon inventors were at work, and light boxes providing 10,000 lux became available and were successfully used by
many with SAD—the SADness washed away with light. Now
new portable models are appearing, the latest weighing no more
than a few pounds.

It is a commentary on the fixation of the corporate mind-set, the deepness of the grooves of old paradigms, that, although they have supported some of the basic research, not a single U.S. manufacturer of lighting fixtures appears to have caught on to the commercial opportunity. The light boxes are made by very small companies, the mom and pop industries, the garage-based enterprises, linked to scientists in this field because they have understood the opportunity and are rising to meet the need. The only major company I have seen with a real understanding of this opportunity is in Japan. Matsushita Electric Works has done the calculations and realized what a remarkable effect bright-light technologies can have on its lighting-fixture business.

SWITCHING ON ALERTNESS AT NIGHT

WHAT LIGHT levels are needed? Although 2,500 lux looks like an appropriate level to correct the sadness of SAD, and higher levels shorten the needed time of exposure, up to 5,000 to 10,000 lux is needed to send you rapidly to a different time zone, as the next chapter will show. But what if we just want to stimulate alertness at night, to flick on the switch in our brains as well as the one on the wall?

At the Institute for Circadian Physiology, we set up a full-scale simulation of an industrial control room in a Human Alertness Research Center (HARC). This was a re-creation of the workplace from where so much of modern industry is operated. The Foxboro Company kindly provided us with the computer consoles and helped create the wall panels of dials and switches to lend authenticity to the stage set. Operating manuals, industry magazines, the soporific sixty-cycle hum of the equipment, and the dim room lighting contributed to the scene, as did the multicolor graphics of the console screens. In HARC we created a simulation so realistic that experienced shift workers told us it felt like the real thing.

Volunteer "operators" working two successive night shifts from 11:00 P.M. to 7:00 A.M. were exposed to light levels of 10, 100, and 1,000 lux. One hundred lux provided no advantage over 10 lux, but 1,000 lux was a different story. The human brain turns out to be much more sensitive to light at night, so one has only to double the brightly lit office level of 500 lux to obtain a significant effect.

That bright light could suppress the waves of sleep that periodically engulf the night-shift worker became clear. The 1,000 lux of light switched off this natural debilitating tendency and, as a result, performance soared. The delay in responding to alarms was reduced by 40 percent. The zombielike stare at the console, the sluggish working of a sleepy brain, were suppressed. Problem-solving functions were enhanced. There was now someone at home.

Even when given the opportunity to nap in the middle of the night shift, most people could no longer do so. Without the supplemental light it was easy enough, and most people spent the shift wishing they could. But with bright light, the subjective sense of alertness was enhanced, the brain switched on.

These findings have since been repeated in many other laboratory studies. The human brain's alertness can be switched on and off by taking a human-centered approach, by acknowledging the conditions the human brain needs to perform. Brighter levels of light push sleep away from the night, especially if days are kept dark, as the next chapter in this story will demonstrate.

My colleagues and I are now introducing this technology into the industrial workplace. Part of the challenge in the real world is that bright-light delivery requires the design of workplaces free of glare and irritating reflection. Nothing is more tiring to nighttime eyes than harsh glare or reflections caused by unthinking use of materials or positioning of lights. The reason we tolerate well 10,000 or even 50,000 lux outdoors is that light is appropriately scattered, and shadows, contrasts, and reflections are minimized. These considerations must be brought indoors to harness the full power of light.

CONTROL OF DAWN AND DUSK

WITH THESE light sources we are still instantly flicking on and off the light. Ignored are the subtleties of the gradual waxing and waning of light at dawn and dusk. But recent evidence suggests these may be very important as we seek to control our lives. The process of waking up from sleep is much influenced by the rising of the light at dawn. Much smaller light levels are needed for the same effects of light if their onset is gradual and specifically timed.

A world can be imagined in which the alarm clock does not emit a rude intrusive sound. Instead the alarm is much more subtle, a much more gentle arousal from the depth of slumber, lifted up by a simulation of dawn's early rays. Body processes nudged into action even before we awake, refreshed for the day ahead.

This is not so far in the future as you might think. The patents are filed, the investment in product development is occurring. Some early versions are already out in the market. We are finally going to see the light.

MINIATURIZING THE LIGHT BOX

LIGHT BOXES and light fixtures are cumbersome things if you want to be free to move around and still benefit from the benefits of bright light indoors or at night. You face the alternative of lighting all your environment or of carrying the local zone of bright light with you.

The answer from inventive minds was the head-mounted light visor. We all have images of miners with lights on their helmets toiling in the dark caverns of the mine, or of ophthalmological surgeons with the light source like a badge of honor perched on their foreheads. But turn the light around to focus on the individual's eyes rather than on what the eyes are seeing, and you have the concept of the light visor. It was first commercially developed by BioBrite for the SAD market, the people who must treat themselves with light every morning of winter to maintain their mood, to ensure their sense that all is well. But it is hard to cope with the duties of running a household, of preparing for work, if one has to be glued to a light fixture. With portability, the light carried with them, the light visor reduced the hassle of everyday life.

We turn in the next chapter to the advantages of this technology for the traveler or shift worker.

Chapter 14

Traveling Through Time

I HAVE SOME good news and I have some bad news. First the good news: our knowledge of how bright light can reset biological clocks means that you now have the ability to travel forward or backward in time. The bad news is that the maximum you can achieve is forward twelve hours or backward twelve hours. Furthermore, you can shift only your body time and not the time of the environment around you. We cannot have you relive the pleasant party of the night before or skip today's presentation at work.

The power we can exert with bright light gives us a fleeting sense of travel through time. We can give you two lunches in a day or two nights of sleep—little use for the time traveler of fiction, but actually of considerable advantage for the citizen of the twenty-four-hour society.

THE PACE OF TRAVEL

THE JET is not a human-centered technology. It serves our desire to transport ourselves to distant places, but it does nothing to help our physiology adjust. We are rudely assaulted with the disparity of body time and local clock hour when we arrive at our destination. Our pragmatic goals are met—travelers are physically transported to their destinations—but their bodies and minds are still in synch with the time zone of their departure.

166

The consequence, depending on the individual's susceptibility, is malaise, dyspepsia, sleep disruption, and fatigue as the body struggles to cope with the suddenly changed rules of dusk and dawn. Meals are presented when the digestive tract is not prepared to receive them, or are not available when hunger pangs rumble. Meetings are held at bedtime, and bed is offered when one is most awake.

Traveling across time zones was not always so hard on the human body. Before air travel, it was not possible to cross more than one or two time zones a day (unless near the North or South Pole, which one can jog around and pass through all the time zones in a few seconds). In contrast, time zones at the equator are more than one thousand miles apart, and even in temperate zones they may be at intervals of six hundred to eight hundred miles, distances beyond the reach of the preaviation traveler.

When I emigrated to North America, after completing medical school in England, I decided to do it as so many had done in the preaviation era: by ocean liner. My trip across thirty-two hundred miles of ocean took nine days. I didn't realize how slow that was until I had been on the ship for a day and we had not even left Penzance behind. I did the calculation 3,200 miles ÷ 9 days = 355 miles per day, or less than 15 miles per hour!

"You mean I have to sit on this boat, tossed up and down by the Atlantic swell, while we chug along at only fifteen miles per hour?" I asked myself. I obviously needed to slow my speeding brain, to adjust to the pace of a bygone era. After a while I got into the rhythm of the ship and enjoyed the time of seemingly suspended animation. We traveled at a pace to which the human body could well adapt. With the style of the past, midnight was delayed an hour every second night, an hour spent dancing to the ship's orchestra. We were well within the capabilities of my biological clock to adjust.

Even when air travel began, humans did not immediately enter the jet-lag age. My father recounts his experience in the 1950s of traveling from London to Kenya by air and taking three days to do it. They traveled at 200 mph in a series of short hops, stopping for meals and overnight hotel accommodation along the way. Lunch in Nice, overnight in Malta, Bengazi for lunch, Wadi Halfa overnight, and so on.

Unfortunately, this is no solution for the hustle-and-bustle world of today. We rarely have the time, although occasionally we should indulge ourselves, to linger across time zones, to travel at the natural pace of our biological clock. Human alertness technology must recognize this fact and ease the human burden of time transition.

LAWS OF LIGHT

THE DISCOVERY of the effect bright light has on resetting the biological clock gives us tremendous power, which must be carefully and responsibly used. The effects of bright light are extremely time-of-day dependent, a feature that is the basis of their action.

As we discussed in Chapter 3, bright light in the middle of the day has little effect, but in the evening gradually increasing sensitivity to bright light is observed. With successive clock hours, greater and greater delays in biological-clock time can be induced. Then past a break point in the middle of the night, typically around 3:00 to 4:00 A.M., around the time that body temperature reaches its nightly nadir, the direction of resetting abruptly reverses. Instead of maximal delays, maximal advances are seen in response to the same bright-light stimulus. The responses then wane in their magnitude the closer the timing of the light signal approaches biological dawn.

In this response, the intensity of the light is important. Below 2,500 lux little in the way of resetting occurs. But as light intensities reach 5,000 lux, 10,000 lux, and beyond, the resetting power is enhanced. The length of exposure also matters; more shift in biological-clock time is achieved with three hours of light than one or two hours.

CREATION OF A PRACTICAL TECHNOLOGY

THE CHALLENGE has been to convert this technology to products and know-how of practical value. No one other than a scientific expert can be expected to carry around the complex function of the human resetting response to light (called the "phase response curve"), nor can people be expected to calculate the precise dose and duration of light needed for their next plane trip. The answer, we realized, was to create a computer software program. My colleague Tom Houpt, who created the code, came up with the name "MidnightSun™"—a name in keeping with the prescriptions it provides.

Using the itinerary from the travel agent and some information about your sleep-wake pattern, together with the software routines that encode, thanks to Tom Houpt's diligent work, every major airport in the world, every time zone, every daylight savings law in every nation of the world—we can calculate the dynamic patterns of light and dark, dawn and dusk, and local time throughout the trip (Figure 14.1). Taking a phase response curve, which is ensconced within its memory, the bright-light protocol for rapid resetting of your biological clock can be calculated and displayed either by graph or table. A program to minimize the impact of jet lag is thus available, tailor-made to your own sleep-wake schedule and a particular trip you will be making on a particular day of the year.

DELIVERING THE LIGHT

SO FAR, SO GOOD. We can tell you how much bright light you need and when you need it. When taking transatlantic trips, you can usually get bright light by walking outside if it is daylight. On other trips—for example, transpacific—you will often find that it is dark outside, so natural light is not on tap, or you are on the plane when you need bright light, or in a hotel room. The standard heavy light boxes are impractical for the traveler. They would be like carrying another briefcase, and one obviously could not plug them in on the plane.

The answer, we decided, is a portable light visor like the ones invented for treating SAD (winter depression) patients. At the time of writing, we are conducting trials of a sleek, lightweight, battery-operated device that can be put on anywhere, including while on a plane. Now working to develop this technology into a device suitable for jet lag, Circadian Technologies has been testing it with corporate executives crisscrossing the globe.

SPRING EQUINOX

Figure 14.1

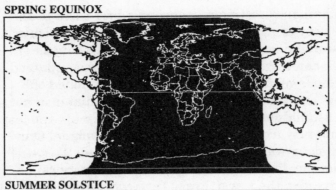

SUMMER SOLSTICE

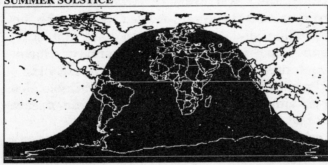

WINTER SOLSTICE

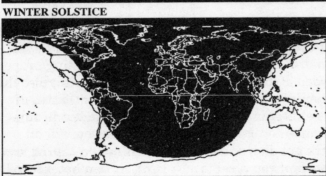

Dr. John Mitchell, the corporate medical director of Amoco, and a colleague were flying to Burma on business and bravely volunteered to be our first guinea pigs. They did not know how brave they were going to have to be. We had them wired up so we could record their biological rhythms, planned the trip and experiment with care, and sent them off on their journey.

Everything went smoothly until they were flying the leg from

Bangkok to Rangoon, when the program called for them to don the light visor. They were sitting on the plane, a Thai Airways Airbus A300, with the visors on when suddenly three hijackers carrying a bomb came rushing up the aisle and commandeered the plane. Mitchell and his colleague immediately ripped off their light visors and our carefully prepared recording instrumentation and stuffed them under their seats. After all, the last thing one wants to be is conspicuous on a hijacked plane! Our experiment, unfortunately, was abruptly terminated. We soon recruited some new volunteers, however—a large party of DuPont executives heading into China—although I first had to assure them there was no correlation between wearing the device and getting hijacked!

The responses of people wearing the light visor device have been very good, and we are refining it for general use. By the time this book is published, the device should be available as a personal tool for physiologically managing the problem of jet lag.

SUPPORTING TECHNOLOGIES

OTHER PARALLEL technologies offer some promise. Evidence is growing that melatonin, the hormone produced by the pineal gland at night, may have biological-clock resetting properties. Doses provided according to preplanned times will reset the biological clock, and the best approach may be coordination of bright light and melatonin at carefully designed temporal relationships to each other.

Vigorous efforts are also under way, including a research and development program sponsored by Upjohn at Stanford, to find other "chronobiotics"—drugs that reset the biological clock. But in claiming this effect, one must be very careful. Some short-acting benzodiazepine sleeping pills have been aggressively promoted as jet-lag cures when they really are not. They will certainly promote sleep on a new schedule or on the plane. However, there is no evidence in humans that at safe doses they reset the biological clock.

This is not to say that sleeping pills may not sometimes be helpful; indeed, they may. Part of the symptomatology of jet lag, but only part, is caused by sleep loss en route. Such drugs can help you sleep better even if they do not help your clock. However, be careful with drugs such as Halcion, and take as small a dose as possible, because it causes severe disorientation in some people. Some countries, such as Great Britain, have banned Halcion.

Another approach has sought to address the psychological time warp caused by jet lag—the sudden shrinking or expansion of the day as one crosses multiple time zones. Ross Mitchell of Acclimator Time Corporation has developed and patented a Jet-Lag watch that provides a gradual transition between time zones. The watch, which runs at a slower or faster pace until you reach your new time zone, is designed to eliminate the sense of unreality of some clock watchers who get perturbed by abrupt changes in time when resetting their watch.

Yet other innovations address the need to sleep and nap at strange hours. For a number of years it has been possible to rent rooms by the hour at hotels, even though these services are rarely used. With increasing sophistication of the traveling public and their awareness of strategies to reset the biological clock, these facilities will be used more often. They will need an upgrade from the present simple rooms to rooms designed to provide recuperative sleep and with lighting systems to ensure the guest wakes up refreshed and has his or her biological clock reset.

Inventive minds are hard at work on such challenges, which must be addressed for human adaptation to the twenty-four-hour society. A broader question, of course, remains: is it necessary to travel so much? Telecommunication technologies and video conferencing are developing so fast that the days of *having to* globe-trot to do business may be turning into a historical anomaly of the twentieth century. Even before the days of teleconferencing, there were those who questioned the need for personal global diplomacy. As former U.S. secretary of state Dean Rusk frequently pointed out, the diplomatic service is a five-hundred-year-old invention designed to make it unnecessary for kings, presidents, prime ministers, and secretaries to be everywhere at once.

Chapter **15**

The Human-Centered Workplace

A WORKPLACE TO be used around the clock should be built to enhance human performance rather than degrade it. The statement seems obvious, yet without realizing it most architects and designers create space that is ill suited to sustain alert and productive human function. Driven by the properties of materials and equipment that are familiar or provide aesthetic "design weight," they fail to meet the most obvious of objectives—aiding the performance of the occupant at all hours of day and night.

The time has come to rethink the design of interior space—to apply human alertness technologies to the workplace and make it suitable for the needs of the human race in a nonstop world. Again, an overhaul of the mind-set is required; old assumptions need to be put aside or reiterated in the light of the new paradigm.

CONFERENCE ROOMS ARE NOT FOR SNOOZING

YOU GATHER together your key employees to impart and process some critical new information on which the future of your company depends. The speaker loads the Kodak carousel with her information-packed visuals and is ready to begin. But first, before the speaker starts her presentation, you undermine the process—you switch off the lights.

173

In the cozy twilight, seated in padded chairs, the audience is lulled into a comfortable state—but this is not the state to ensure optimal communication. Hold the conference at a time of day when the barriers to sleep are lowered—just after lunch or late at night—and a significant proportion of audience members will lose their capability for rapt attention. In other words, they will nod off.

I have participated in or given countless briefings in the early afternoon, postlunch-dip period, during which eyelids were drooping in the key participants—even when the presentation was as lively and interactive as possible. The more plush the conference room—usually associated with the importance of the audience—the more likely is the environment designed to defeat the communicative purpose. The combined value of the audience's time makes this a very expensive mistake. A conference involving, say, twelve key executives with an average salary of $200,000 per year, plus a couple of $3,000 per day consultants, costs about $2,000 per hour for the time alone, let alone the travel time to attend the meeting and the cost of the misinformed decisions that may be made by a sleepy audience.

In these conference rooms, I keep the lights on as much as possible (dimmer switches and accent lighting help), even to the point of reducing the sharpness of my slide presentation. I also try to make the presentation lively. But this is just a superficial quick fix to a ridiculous situation, especially considering that the room was designed for the very purpose for which I am using it.

Let us think about where the workplace problems lie. The best approach is to run down our checklist of the nine switches of alertness and figure out under whose responsibility each switch lies.

Alertness Switch	*Responsibility*
1. Sense of Danger, Interest, or Opportunity	Speaker
2. Muscular Activity	Meeting planner
3. Time of Day	Meeting planner
4. Sleep Bank Balance	Participants
5. Ingested Nutrients and Chemicals	Participants

6. Environmental Light	Architect/designer
7. Environmental Temperature and Humidity	Architect/designer
8. Environmental Sound	Architect/designer
9. Environmental Aroma	Architect/designer

Some of the nine areas, covered in Chapter 4, are the responsibility of the speaker, the meeting planner, or the participants. However, light, temperature, sound, and aroma of the workplace fall fairly and squarely in the architect/designer's domain. We will now deal with each of these in turn.

• Light

The most obvious problem is the dim—and sometimes pitch-dark—lighting when presentations are given. The reason is the technology chosen to provide visual graphics—either a Kodak carousel slide projector or an overhead viewgraph projector—require the room to be dark. Yet technologies are available to show visuals on screens that can be seen even in sunlight.

Consider the large screens used in sports stadiums made out of composite color bulbs that create an image visible in the brightest of sunlight. This technology could be converted into the visual screen for a conference room that could be bathed with artificial light, daylight, or even direct sunlight during presentations.

Chapter 13 addressed the powerful effect room lighting can have on human alertness, especially at night. Clearly, these bright-light systems need to incorporated into workplaces that are used at night. It requires a radical rethinking of materials and surfaces to ensure that glare does not irritate tired eyes.

Another emerging science is the color of light and the effects on human performance and mood. Not yet fully documented and under intense development in research and development labs, it offers further capacities for affecting moods.

• Temperature

When comfort is the only goal, alertness may be lost. The idea should be to maintain a comfortable level of temperature for the lower body to reduce distraction, but to provide cooler air at the level of the head. The face is especially sensitive to cooler temperature, and this can help the alertness of the brain. Consider what you do when wishing to freshen up and alert yourself—you splash cold water on the face, which has a marked stimulant effect. Design efforts need to be made to provide stratified temperature system to counterbalance the normal temperature gradients in a room where warm air naturally rises. We need to invert that law of physics to pay respect to the law of alertness.

• Sound

A proper balance needs to be achieved between the soothing effects of humming or white noise and the distraction caused by more discrete noise. Clearly, nothing should be done to distract the audience or the speaker.

• Aroma

A new technology is emerging that uses different aromas piped in by aromatic systems. Shimizu in Japan, for example, has pioneered such ventilation engineering systems. Certain aromas are claimed to stimulate alertness; others induce a more relaxed state. This science is still young, however.

STRATEGIES FOR SUSTAINING ALERTNESS

WHEN PEOPLE are working very long hours to complete a deal, finish a critical project for a deadline, or manage an emergency situation, the facilities described for sustaining alertness may not be enough. For people working flat-out around the clock, other strategies must be used.

To determine what facilities the workplace of the future may need to provide, let us examine the correct way to deploy scarce human resources under the most demanding circumstances such as earthquake relief or rescue operations after an airplane crash in remote terrain. Do you push people until they drop from exhaustion, or is there some other way?

The answer lies in how long you expect the emergency to last. If it will be finished within twenty-four hours after the emergency team's last regular sleep period, the most effective strategy is to push through day and night until the task is done—but allow time out for a strategic nap prior to the individual's normal sleep time. The timing is critical, for tired workers who nap when they are most sleepy in the wee hours of the morning will suffer from the extreme grogginess of sleep inertia and may be worse off than if they had not napped at all. In contrast, the prophylactic evening nap will sustain them through the dead of night, even though they may not feel very sleepy at the time they take the nap.

An emergency expected to take longer than twenty-four hours of continuous effort requires a different strategy. Pushing people until they collapse from fatigue will render them ineffective and incompetent, and more will be lost than gained.

An effective strategy to cope with these longer-term emergencies has emerged out of long-distance solo ocean racing. Here the need for sustained attention twenty-four hours a day for weeks on end is critical. The time spent sleeping is time a sailor risks being blown off course or does not take full advantage of the prevailing winds. My colleague Claudio Stampi, a twice around-the-world ocean sailor who is also a leading sleep physiologist, noticed that the most successful racers adopted a pattern of repeated short catnaps. Indeed, there was a clear correlation between who was winning the races and how much they broke their sleep into short naps.

His explorations into the patterns of ultrashort sleep bouts or "polyphasic sleep" have yielded some fascinating insights on the nature of human sleep and also some highly practical applications for emergency situations. As discussed in Chapter 3, most animals actually catnap around the clock rather than adopting the unusual human habit of taking all our sleep in a single consolidated episode at night. But we still have vestiges of that age-old pattern of behavior, and with some training we can return to that more primitive state.

Stampi has turned into a science a secret that some of the most productive people throughout history have figured out for themselves. Leonardo da Vinci, the original Renaissance man, has been said to have cut down his sleep need to two hours a day by breaking it into multiple catnaps. Thomas Edison, who worked at a nonstop pace to amass more significant patents than any other inventor, used catnaps as a way of keeping going throughout day and night.

Such procedures can be developed for use in emergency situations. Tests lasting as long as two months have shown that it is possible to take seven- to eight-hour-a-night sleepers and convert them into two- to three-hour sleepers and still maintain high levels of performance around the clock for a period of several weeks. In these strategic rethinks of how humans should sleep when the ultimate in performance is required, Stampi has shown that a twenty- to thirty-minute nap every four hours around the clock can keep a person sustained so that twenty-one to twenty-two hours of productive work can be accomplished per day. When naps are extended for longer than twenty or thirty minutes, one slips into too deep a sleep stage and tends to wake up in a groggy state. Thus naps should be shorter and more frequent rather than longer and fewer.

One of the most remarkable studies that proved the effectiveness of sustaining people with naps on long, drawn-out emergencies was recently conducted by the United States Army. When I learned that the army was doing a study on productivity during warfare, I was more than a little curious about what measure it would use. The measure turned out to be how many shells per hour a gunnery crew could land on a target. Two gunnery crews were given the task over a three-day period of shooting the maximum number of shells into a target area. One group blitzed through night and day, continuously firing away without taking a break for sleep. The other took a more paced approach, stopping at intervals for a nap before continuing its bombardment.

The old fable of the tortoise and the hare might inspire you to conclude that the napping hare would lose the race to the continuously plodding tortoise. However, the fable was about a relatively short race, not a continuous exercise lasting several days. In the army study, the nonstop crew surged ahead in its on-target hits over the first twenty-four hours as the other crew took breaks to nap. But by the second day, the napping crew had caught up and overtaken the blitzers, because the nappers remained sharp and effective, whereas the accuracy and energy of the continuous blitzers lagged.

It is all a question of understanding the design specs of the human machine. Efforts to ignore the truths about human physiology and try to dispense with something as "sissy" as sleep will result in inefficiencies and failures. The commander on the battlefield, the leader of a rescue mission, the captain of a malfunctioning spaceship, will not achieve the ultimate in human performance without acknowledgment of how one best sustains the performance of the human machine.

FACILITIES FOR REST

THE LESSONS learned from the most demanding work-places have relevance for more ordinary around-the-clock operations. If you are going to prevent inadvertent snoozing at times and places that are inappropriate, the enlightened work-place of the future will need to provide facilities suitable for recuperation.

The technology of short naps can be usefully applied, provided that there are suitable facilities and a workplace culture that supports such activity. In most American workplaces, napping is frowned on at best, and at worst a matter for severe disciplinary action. In Mediterranean and Latin American cultures, however, a siesta hour is the norm. Napping is to a large extent a learned behavior, and cultural attitudes and the climate determine behavior. However, some people find napping easier to achieve than others.

An approach developed in Japan by Matshusita shows the way of the future. A specially designed Rest and Relaxation Chair lulls you to a relaxed state as it automatically reclines and massages your back. Then, as you sink into the early stages of drowsiness, it times your entry into the twilight of consciousness before switching at a predefined time to a wake-up mode in which the chair back is raised, bright lights are switched on, cool air is blown across your face, and you, refreshed, are ready to go back to work.

I have tried out early prototypes of these chairs, and for me the subjective sense of renewed vigor and alertness is clearly achieved. We are now working with Matsushita to develop objective assessments of the efficacy of this equipment, but it is clear that this technology is coming to the workplace of the future.

THE TWENTY-FIRST-CENTURY BEDROOM

PERHAPS WE will be able to keep many of our workers from needing to nap at work by improving their sleep at home. Some fascinating new technologies are emerging to help us with this goal.

The bed itself, while looking very much like the beds to which we are accustomed today, will have much ingenuity and technology below the surface. The intelligent bed will contain sensors that can detect the pattern of your heartbeat and respiratory movements, and process this information so that the states of sleep or wakefulness can be determined. Because the bed knows whether you are awake or asleep, and can even determine the quality of your sleep, a door is opened to creating a feedback loop so that appropriate corrective action can be taken. Other technologies will enhance the onset of slumber and reduce the time spent restlessly awake. The environment of the bedroom will also be modulated to serve the overriding goal. This bed thus becomes a human-centered tool to increase the quality of our rest.

The art and science of creating such a bed are still in the primitive stages, but progress is being made.

The technology of arousal will also be developed with the same human-centered approach. If there is a symbol of our machine-centered world, it is the alarm clock, which rudely, without preparation, awakes us from slumber. New approaches have already been devised that control the lighting in the bedroom to mimic the natural dawn. Rather than an abrupt awakening, the level of illumination starts rising thirty minutes or even an hour before the desired wake-up time. Arousal is smooth and less abrupt, initial studies show. Thus, by understanding the natural patterns of dawn that the human biological clock saw for the first million years of its existence, we can create a resonance with built-in mechanisms of our brains.

Still other efforts will seek to understand further the dynamics of two people asleep in a bed together. The strange unconscious dance, so well captured on film by Dr. Alan Hobson of Harvard Medical School, needs thought to conceive of ways to adjust a bedroom to the needs of two people who may be physiologically quite different. Indeed, it often seems that owls marry larks, and long sleepers marry short sleepers, so opposites do attract. The warmth or lighting on one side of the bed need not be the same as on the other; it is important to acknowledge individual needs and respond to them.

All this technology may sound too complicated, and may sometimes be unnecessary if people can get sufficient control over their lifestyles, as we will discuss in Chapter 18. But for the many people forced to work and sleep at unusual hours, such bedroom technology may at last give them control over their rhythms of sleep and wakefulness.

RETHINKING ASSUMPTIONS

A WHOLE NEW way of thinking is needed, backed by objective scientific research, about what qualities of a workplace or living space promote creativity, productivity, alertness, and safety. Should we be bottling up the natural characteristics of the outdoors when we build interior space? Or are there ingredients of the outdoors that induce moods and states not conducive to work and concentration? If you re-create for me the ambience and smells of an English meadow on a warm summer afternoon, you will lull me into blissful slumber. What I need is a brisk cool autumnal day without too much sunlight to provide the appropriate ambience for work.

Today architects and designers operate from considerations of only cost and aesthetics, with a smattering of the primitive knowledge found in most human factor manuals. These manuals need to be rewritten to encompass the needs of the daytime and the nighttime worker. They should address seriously the problem of glare in the workplace, the main factor that inhibits the use of light intensities appropriate to optimal human function.

It may be necessary to develop programs for interior space, as Shimizu Corporation in Japan is promoting, so that they can be adjusted to the purpose at any particular time. Resembling a stage set with multiple light sources, sounds, smells, visual effects, and air flows, these high-tech workplaces of the future for the first time address seriously the stimuli that humans need to support their fully effective function. Is there perhaps a set of environmental conditions that promotes brainstorming creativity, another for quiet contemplative thought, and another for communication? This is the claim, and it is worth studying carefully if we are to develop a human-centered world.

IV

TRANS-FORMATION INTO A HUMAN-CENTERED SOCIETY

Chapter 16

National Challenges in a Nonstop World

IN OUR RAPIDLY coalescing global economy, with trade barriers falling and competence competing worldwide, concern is rising in each nation of the world about its ability to compete. Successful competition means delivery of the highest quality for value; it means harnessing the resources of each nation in the most effective way. Above all, it means that human ability and performance must be at their peak.

Yet, as we have seen, the features that make our economy global and nonstop also induce a pervasive human fatigue that saps creativity, performance, and drive to excel. The increasing reliance on automation and on twenty-four-hour-a-day production pulls the most critical employees in the producing economy into a world for which their bodies were not designed. The effects of this societal transformation on productivity, quality, and competitiveness are so important that nations, corporations, and individuals who solve these problems in the global economy have a tremendous advantage to gain.

It is time to look into the future, to look at the challenges that lie ahead, if we are to make a successful transformation into a human-centered twenty-four-hour society. Let us look first at the challenges that various nations and cultures face, and then in subsequent chapters at strategies that corporations and individuals must adopt to succeed and flourish.

THE AMERICAN ADVANTAGE

EACH NATION has its own unique advantages and disadvantages as it moves into the global twenty-four hour society. From my experience of visiting various countries around the world and acting as consultant to some of their industrial operations, it has become clear to me that the United States possesses some decided advantages.

First, a comprehensive twenty-four hour infrastructure has developed. In most major cities, supermarkets, pharmacies, entertainment and recreational facilities are open around the clock, six or seven days a week. It is possible to work on any type of schedule and still shop on your own schedule and obtain the other services you require.

Second, US managers over the past ten years have become increasingly aware of the problems of human fatigue in round-the-clock operations and are quite active in seeking solutions.

Third, many people who work round-the-clock shifts have recognized the importance of adapting their homes and bedrooms so they can optimize sleep and recuperation at any hour of day or night. American health care costs and legal liability costs may be out of control, but for operating an efficient twenty-four hour society, the USA has a clear advantage.

Japan and the Pacific Rim Orient are tremendously challenged when it comes to operating as effectively at night as during the day. The Japanese custom of housing an entire family in one or two rooms—switching them between bedrooms at night and living areas during the day—makes it extraordinarily difficult for the night worker to get good-quality daytime sleep at home. The thin walls that sounds penetrate, and the limited space for family life driven by an incredible shortage of land and sky-high real estate costs, conspire to reduce the effectiveness of the night worker by depriving him of the opportunity for good-quality sleep.

Riding in the commuter trains in any Japanese city, one is surrounded by sleeping people, whose capacity to sleep while hanging onto straps or squashed tightly into seats is remarkable. This is a phenomenon not seen, except for the occasional individual, in any city in North America or Europe. Why are the

Japanese daytime-working commuters sleeping? The Japanese may have a national debt in their sleep bank—their fatigue may not be adequately recuperated even by nighttime sleep.

It is probably much worse for the shift worker, and there are signs that this is the case. The Japanese are much more openly supportive of workers taking breaks to sleep on the night shift; for example, even in nuclear power plants this is a well-accepted custom. Under carefully controlled circumstances, as mentioned earlier, this can be an effective strategy. But in Japan's case it does show the pressure to sleep is strongly felt.

Many facilities in Japan and elsewhere in the Orient provide special sleeping quarters at the work site to address this problem. Some have dormitories and beds in the factories—a strange sight to American or European eyes. In contrast, Americans and Europeans have much more spacious homes, where shift workers often can construct special sleeping quarters for themselves and isolate themselves from family and the interruptions of the daytime world.

The British and other leading European nations, however, have a handicap of a different nature. It is commonly assumed in Europe that shift work is a bad thing and should be curbed by strong, intrusive regulations promulgated by unions and government alike. European occupational health scientists have spent considerable time cataloging the health and error problems of shift work and have contributed to the clamour for tighter regulation. They have spent less effort looking for practical, workable solutions or pioneering new technologies that will help people make the inevitable adjustment to a twenty-four hour society. They have instead taken steps to introduce regulations that can only stifle European competitiveness in a twenty-four hour world.

For example, regulators in European countries have taken steps to ban twelve-hour shifts and reduce night work as much as possible, both seemingly desirable but ill-informed goals. Based on skimpy or even absent data, twelve-hour shifts have been assumed to result in more fatigue and stress, more errors and human failure. However, evidence from the many workplaces in the United States that have converted to twelve-hour shifts suggests that this is not the case. Twelve-hour shifts provide twice as many days off work to recover than most eight-hour work schedules. This allows much more time for recuperative sleep and improves the quality of family and social life for the shift worker and his or her family.

Wait, you might ask, how can a person working for twelve hours not be more fatigued than someone working for eight hours? Exhaustive studies by our research institute examining twelve-hour and eight-hour shift operations has shown that sleep quality and alertness appear at least as good on twelve-hour shifts as on eight-hour shifts. The reason is that alertness and the lack of a sleep debt are the major contributors to performance on the job, not the precise number of daily work hours—within limits. I do not argue for fourteen- or sixteen-hour shifts because then consecutive hours on the job may take a toll; also, some highly demanding jobs that involve much physical labor may also not be feasible for expansion to twelve-hour shifts.

This is not to say that twelve-hour shifts are free of problems. The transition to the new schedule must be carefully designed, and supervisory and management problems must be addressed, as must various issues of coverage, benefits, and the like. However, major improvements in efficiency and communication result from having only two crews per day each meeting each other on the way in and out, and only two versus three crew turnovers per day, reducing by one-third the number of shift changes, when low productivity and errors commonly occur. By banning such twelve-hour shifts, Europeans are losing flexibility in finding the best solutions to human performance problems in shift work; opportunities to make real gains in productivity are swept away by the regulator's pen.

Similar unnecessary problems are also created by the regulatory attempts to minimize night work. In a world where capital is limited and global competition is intense, driving down costs and improving return on assets or investment (ROA or ROI) is critical. The lowest costs and the highest ROA are achieved when a capital-intense facility is used 168 hours per week, rather than building two facilities and using them 84 hours per week or, even worse, four plants and running them only 40 or 42 hours (that is, nine to five, five days per week). Those nine-to-five facilities may provide a more "civilized" life for their employees, but life is not civilized if you cannot afford the investment to create those jobs, to give those prospective employees a chance.

The United States thus has a substantial advantage in terms of homes that can accommodate a night worker's strange hours and in the restraint on regulation that is the American way. Indeed, a major limitation on twelve-hour shifts—the provision in the Walsh-Healey Act that overtime must be paid after eight hours of work in a day—was swept away in recent years, thus making it cost-effective to vary the work hours in a week, necessitated by the arithmetic fact that twelve-hour shifts do not evenly divide into a forty-hour workweek.

Great Britain has for centuries enjoyed the status of a global power and London is still one of the world's most important financial centers. The opportunity exists for continued British leadership in the evolving non-stop global village. But this will only happen if Britain embraces the challenge of addressing human limitations in a twenty-four hour world. The economic and technological benefits to be gained are enormous, provided the human side of the equation is managed with creativity rather than inflexible regulation.

NATIONAL STRATEGIES FOR THE TWENTY-FIRST CENTURY

THE UNITED STATES' economic challenge of the twenty-first century is to build on this advantage in lowered costs of capital investment, and lack of regulatory restrictions, and to create a truly successful and humane twenty-four-hour society. This will require all the energy and creativity the country can muster to create a human-centered world.

The challenge for Europe is to take careful stock of where regulatory zeal is taking it. The twenty-four-hour society cannot be legislated out of existence; Europe has to be part of the global village for its economy to flourish and benefit. The creation of good-quality manufacturing jobs can be assured by ensuring that it is feasible to run around-the-clock operations cost-effectively in Europe. As a Britisher myself, I am fully aware of the importance of maintaining a cultural heritage and way of life. But we must face the twenty-four-hour society challenge as well.

Japan's surest advantage is that its major companies and its engineers have already started to embrace in a comprehensive way the challenge of creating a human-centered world. Perhaps because Americans and Europeans have not yet developed the conceptual vocabulary of a human-centered world, they have failed to read the message coming through loud and clear from Japanese companies about the need to design machines sensitive to the needs of the human body and mind. Their innovative capacity on this frontier is outstanding, although they have yet to completely fit the social fabric to the technological dream.

In our quick spin around the globe, we must not forget the developing world, where infrastructure support for twenty-four-hour operations is much weaker. State-of-the-art factories can be built and staffed, but the nighttime accident that killed thousands at Bhopal has shown the need for greater sophistication and oversight. The challenge in such environments is to build a local network to provide the services and support—sleeping quarters, food availability, and training—that would otherwise be provided by society itself in the industrialized world.

Chapter 17

Human-Centered Management: *The Strategic Business Advantage of Placing People First*

COMPANIES WILL set themselves apart in the twenty-first century by how well they optimize the human-centeredness of their technology. The slogan "Our people are our most important asset" will take on a deeper meaning, for it is the alertness and performance of a corporation's people that determines how effectively the corporation uses all its other resources. Human assets must be enhanced and nurtured twenty-four hours a day around the globe by applying human alertness technology throughout the organization, and by educating managers, engineers, and employees in the principles of human-centered management (HCM). HCM is the process by which the company's technology and facilities are made fully supportive of human physiological and psychological needs, so that human creativity, efficiency, safety, and health can flourish while the other benefits of technical innovation are fully exploited to the corporation's advantage.

Accomplishing this transformation into HCM will be no easy feat. It cannot be left to intrapreneurial efforts by isolated individuals or groups within the organization—although such bottom-up initiatives will be important. Rather, it will be essential for corporate leaders to take charge *strategically,* because significant organizational changes will need to be made. A serious commitment by the management team will be required to make

191

human-centered management one of the fundamental values of the organization, on a par with other essential corporate values such as the drive for quality, safety, and customer service.

To date most companies have addressed this issue only *tactically*—making changes in response to a particular problem at a particular site. Employee problems with the shiftwork schedule lead to a search for a healthier, less human-disruptive solution. A disciplinary problem with employees falling asleep on the job leads to improved training on how to manage a shiftwork lifestyle at that site. And in the more extreme case, a chemical leak, a plant explosion, or a serious injury will precipitate immediate reactive intervention that results in the latest set of additional standard operating procedures.

As documented in Part II of this book, striking improvements in productivity, health, and safety have been achieved where human alertness technology and management principles have been applied tactically, but, with rare exceptions, no fundamental change has yet been achieved in the corporation as a whole. Until a company becomes proactive in the pursuit of human-centered management, these improvements will be isolated and fall far short of the available potential. Man-machine interfaces may still be purchased at the same facility using machine-centered decision making that inadvertently induces operator drowsiness, and corporate policies may be instigated by well-meaning senior management that actually deprive employees of ways to stay alert at work.

THE INTRODUCTION OF HUMAN-CENTERED MANAGEMENT

THE TRANSITION to HCM must be made systemically by a carefully designed process so that the concept becomes not a passing fad but a fundamental, ongoing corporate commitment that is thoroughly integrated into a company's operating culture. To achieve this, a corporation should undertake the following steps to develop, install, and institutionalize a continuous HCM process:

- **R**eevaluate human asset potential and priorities.
- **E**stablish management commitment and support.
- **D**efine HCM as an integral part of company philosophy and mission.
- **A**ssess current risks, liabilities, and hazardous exposures.
- **L**aunch appropriate change initiatives to reduce exposures and capture performance-improvement opportunities.
- **E**ducate and provide ongoing support and training for key personnel at all levels.
- **R**eport results and measure performance to plan.
- **T**ranslate HCM into a continuous, institutionalized process for improving overall productivity, quality, and safety performance.

HCM cannot be institutionalized overnight. The process requires a broad education and commitment to modify established procedures and concepts. It must be a deliberate, step-by-step process, building support along the way throughout the organization.

The HCM institutionalization process is outlined below in more detail.

REEVALUATE HUMAN ASSET POTENTIAL AND PRIORITIES

ESTABLISHMENT of HCM requires a fundamental rethinking of how human assets and technology assets are viewed by the company. A paradigm shift in thinking about corporate priorities, from machine-centered priorities toward human-centered priorities, is needed. Rather than abandoning the advantages of technology, this instead means making sure the technology enhances rather than degrades human performance. The implications of HCM for the other aspects of the corporate mission also need to be explored, and the potential gains from enhancing the performance, productivity, safety, and health of the work force must be assessed.

ESTABLISH MANAGEMENT COMMITMENT AND SUPPORT

HCM MUST be incorporated into the vision of the company as it prepares itself for the twenty-first century. This requires the development of a management commitment to the HCM process, to the technological and management implications of putting people first.

This commitment to HCM by all levels of management must start at the top echelons of the corporation. For HCM to be successfully implemented, it must be introduced and embraced by senior management as well as visibly supported and continuously reinforced. Changes in job description, supervisory and management structure, purchasing decisions, corporate policies, and much more may be required. These will be impossible to achieve without a strong management commitment and clear direction and leadership from the top. As the old ham and eggs analogy goes, the chicken was only involved in the preparation of your breakfast—but the hog was committed.

DEFINE HCM AS AN INTEGRAL PART OF COMPANY PHILOSOPHY AND MISSION

ONCE THE management commitment has been made, the process of HCM introduction needs to be clearly communicated throughout the company. To accomplish this, company philosophy and mission must be carefully defined to support the HCM commitment.

The idea that people come first—that people requirements, not machine requirements, have priority—must be established as a corporate value in the same way that commitment to quality, safety, and customer service is incorporated into the mission statement.

ASSESS CURRENT RISKS, LIABILITIES, AND HAZARDOUS EXPOSURES

WITH AN appreciation of the goals and the concepts of HCM, and the management commitment to HCM communicated throughout the organization, the risks, liabilities, and hazardous exposures can be assessed. The aim is to identify the components of the organization that are most sensitive to the risks of human fatigue or machine-centered decision making. Through a careful process of evaluating facilities, man-machine interfaces, employee personnel policies, engineering decision making, line management structure, and knowledge levels of managers and engineers, the symptoms and the causes of human performance failures in the organization can be identified.

As part of this process, it is vital to obtain the candid input of the frontline troops in around-the-clock operations on the factors that compromise their alertness and performance. Safe feedback mechanisms, whether they be confidential surveys, anonymous incident reporting procedures, or risk-free focus groups, are essential to learning what goes on in the dark recesses of an organization in the middle of the night, and at other times when human alertness and performance are compromised to the potential detriment of the company.

LAUNCH APPROPRIATE CHANGE INITIATIVES TO REDUCE EXPOSURES AND CAPTURE PERFORMANCE-IMPROVEMENT OPPORTUNITIES

THE HCM assessment of an operation will enable one to identify the priority issues and opportunities to be addressed based on the impact they have for reducing risk or improving performance in the company. A series of specific benchmarks can then be defined that serve as the initial specific aims of the HCM process.

Specific programs, policy changes, or organizational changes are formulated and introduced at this point, commencing an ongoing process of transitioning toward and optimizing HCM that will eventually become institutionalized (see below) in the organization.

EDUCATE AND PROVIDE ONGOING SUPPORT AND TRAINING FOR PERSONNEL AT ALL LEVELS

SYSTEMIC CHANGE of an organization's mind-set and operating procedures requires that people at all levels, from senior management to around-the-clock frontline troops, not only become aware of human alertness technology and human-centered thinking but also embrace and use its tenets on a daily basis. Substantially more than information and awareness need to be imparted. An effective education and training process will encourage the creativity and resourcefulness of the innovators at all levels in the organization, who need to find ways to apply this alertness technology and this mind-set to the myriad aspects of around-the-clock operations. It will also fundamentally change the attitudes, behavior, and application skills needed to internalize the process. This may be facilitated by books, seminars, training programs, and hands-on consulting interventions tailored to needs of each manager and employee. This training and support must be an ongoing process so managers and employees can keep abreast of new information and its implications.

REPORT RESULTS AND MEASURE PERFORMANCE TO PLAN

IT IS IMPORTANT to set up mechanisms to obtain the data needed to determine if the benchmarks and overall goals and objectives set for HCM are being achieved. With appropriate measures in place, ongoing performance can be assessed and necessary corrections and modifications made as needed to ensure compliance and achievement of the full benefits of the HCM process.

TRANSLATE HCM INTO A CONTINUOUS, INSTITUTIONALIZED PROCESS FOR IMPROVING OVERALL PRODUCTIVITY, QUALITY, AND SAFETY PERFORMANCE

HCM IS AN ongoing commitment that requires continuous management processes in place to sustain and continuously improve it. The goal is to institutionalize HCM so that it is integrated fully into the decision-making and thought processes of the company. Purchasing decisions, payroll policies, supervisory and management structures, training procedures, research and development initiatives, and much more all need to be cognizant of the principles of HCM.

A company that has fully integrated HCM will create work spaces that promote optimal human performance, man-machine interfaces that support rather than hinder the operator, schedules that help people work at their best, and safety procedures that allow for human frailty.

Chapter 18

Developing Respect for the Human Machine

IT IS 5:30 A.M. in Washington, D.C., but President George Bush of the United States has already put in a long day in Tokyo, Japan. Suddenly, under the unforgiving eye of the TV cameras, he vomits, collapses, and slides under the table at a banquet with the Japanese prime minister, Kiichi Miyazawa, where Bush is the guest of honor. The news reverberates around the airwaves of the world, and the London stock exchange index dips in concern about the impact of the news. Back at home, political hopefuls instantly reevaluate their chances in an upcoming election year, and hasty conferences are called to decide how to get the briefcase with nuclear missile codes back to Vice President Quayle should the worst fears come true.

Bush's enthusiasm for sports had led him to play a couple of sets of tennis with the emperor of Japan—in the middle of the night on Washington time. His enthusiasm for international diplomacy and reelection had led him to undertake a grueling schedule of ten punishing sixteen-hour days on the back side of the clock. Despite extreme fatigue, he had trouble sleeping during the Japanese nights and took a Halcion sleeping pill. His biological clock was still set somewhere in mid-Pacific and had not yet joined him in Japan. He became just one more victim of the human drive to reach beyond our physiological capabilities, to live and work in the global village as though it were a village of a bygone era.

198

WHEN A PRESIDENT becomes sick and embarrasses his hosts because he travels too far too fast, when a worldwide investor or businessperson makes dumb decisions because she is fatigued by around-the-clock phone calls, when a vacationer's holiday is ruined by jet lag—all the potential benefits from the technology of the twenty-four-hour society are immediately erased.

The richness of opportunities in the global village makes a greater range of choices available than ever before. But the capacity to abuse our bodies is greater than ever. Choices without wisdom and restraint can mean disaster and can undo the very benefits which the technology provides.

UNDERSTANDING HUMAN LIMITS

FUNDAMENTAL to living happily, healthfully, and productively in our newly emerging society is a strong sense of the capabilities of the human machine. This is true for ourselves personally and for our employees and colleagues. Whenever we make business decisions that concern people, and virtually all of them do in some way, we must be fully conscious of human limitations. Just because something *can* be done does not mean it *should* be done. Keeping daytime employees working late into the night to meet a deadline can be an exciting challenge—whether for Wall Street lawyers closing a deal or engineers developing a product for a deadline. But it may not be a wise choice if avoidable errors creep into the result.

I will never say never, for sometimes extraordinary efforts are required to accomplish a critical task. But the general rule is that it is unwise to burn the candle at both ends. If you are going to work around the clock, staff your operation accordingly. If you are going to operate a global business, have respect for your employees in each time zone. Make sure that people are trained and fully adapted to work at night or on different time zones. Design facilities and develop procedures to ensure your employees' alertness and effectiveness all times they are on duty. Designate off-duty periods and rest breaks with the needs of the human body in mind. Bite the bullet and understand what you are trying to achieve. It is very enlightened self-interest.

Such advice is especially apropos for presidents, prime ministers, and corporate leaders. With the power they can exert over their own schedule and the facilities at their disposal, one would think they would make wise choices concerning their own physiological limitations. But that power is often abused by them and by their handlers in devising schedules that impair their ability to function.

Unfortunately for us, technology is moving ahead so fast and providing us with so much competence that it too readily makes us victims of ourselves. Competence without wisdom is a deadly combination, a risk that must be consciously contained.

SQUARE PEGS IN ROUND HOLES

WITHIN THE complex structure of the twenty-four-hour society some people do not fit in certain slots. We must each understand the slots in which we best fit and carve out a lifestyle that is human-centered to our own personal humanity.

When we describe human physiology or psychology, the words *normal* and *abnormal* are commonly bandied about. However, there is no such thing as "normal" when one is describing a human being. We are all "abnormal" or unique in certain ways. The various features of our uniqueness fit us well to certain lifestyles; to other ways of living, they make us ill suited.

Early morning types are unwise to take a job that forces them to stay up late; unrepentant night owls should not deliver the morning paper. Long sleepers should avoid twelve-hour work shifts; insomniacs or low-sleepiness-threshold individuals should not perform long-distance driving jobs. Each of us must seek a niche that suits our own physiological individuality.

Knowing your own physiology and psychology and your own limits is essential for identifying your own special niche in the twenty-four-hour society. You will never succeed where you are fundamentally ill suited. Some "misfits" may climb by sheer willpower to the top of their profession, but the emotional and medical costs will take a toll if their biology is incompatible with their chosen career.

Some tests are medical. Certainly, there are well-defined counterindications to an around-the-clock lifestyle, including a history of coronary artery disease, diabetes, epilepsy, or sleep pathology. Company physicals should take care to preclude those who will suffer unduly from rotating shifts and become "cost centers" on the company health plan. There are also standardized tests that determine the characteristics of the biological clock—that identify the extreme early-morning types and the extreme owls. Rigidity of sleeping patterns, even on weekends, and difficulty with Monday morning blues are characteristics that must be acknowledged and respected.

Some niches may become ill suited with advancing age. Many people who adapt well to rotating hours of sleep and work in their twenties and thirties find the rotating lifestyle problematic when fundamental sleep patterns change in their early forties. Shift work may be like an athletic career for some—relatively well paid when one is young, and then an early retirement. In general, there is more propensity for nighttime insomnia and daytime napping with advancing age. As the changes become more extreme, more demanding lifestyles may grow beyond one's reach. Jet lag, too, tends to bother us more as we get older.

But each individual is different, and knowing oneself and one's personal limitations is the key. This is the start to making wise choices and the basis for the advice that follows for living in a nonstop world.

THE INFORMATION BOMBARDMENT

THE CITIZEN of the global village is constantly bombarded with information and besieged by a surfeit of opportunity and concern. Crises, fears, excitement, and entertainment are offered continuously around the clock from multimedia sources, resulting in information overload. Ideas flood in from all over the planet—a technical innovation from Japan, a research study in Australia, a discovery in France. In every field there is no shortage of ideas, just the frustration of having to discard and forget so many.

In this enormous global village, how can the businessperson set reasonable goals when the opportunities and stakes are so huge? Aren't the fortunes achievable by Rockefellers and Vanderbilts in previous eras small potatoes when the whole world is your oyster? Here are some guidelines.

DOING BUSINESS IN THE GLOBAL VILLAGE

WE ARE ALREADY citizens of the global village; the barriers have fallen. The nationality of origin of the products in our stores and homes is of amazing variety. I once assembled a bicycle out of a box—not my choice, but the way it came—and the individual parts came from different countries. The wheels were from France, gears from England, frame from Hong Kong, and so on. This international bicycle never worked very well, but it was probably due more to my lack of mechanical skills than to the incompatibility of nations.

Economic imperatives force businesses to become international to plumb the strengths and the economics of different nations. Efficiencies of operation demand organization by function rather than along national lines. David Morton, the chairman and CEO of Alcan, the major aluminum producer, recently visited us in Boston. He talked about Alcan's consolidation of operations so that all rolling mills worldwide were under one manager, all smelting plants under another, wherever in the world they may be.

The improvement in coordination, the managing of worldwide gluts or demands, the advantage taken of local economic conditions—all argue strongly for such an approach. Yet we must examine how we help every senior manager run a global enterprise and make decisions night and day.

At the same time, the chief financial officers of the world's corporations must keep track of the ceaseless ebb and flow of financial markets around the world. Stock markets, exchange rates, and short-term interest rates are all fluctuating nonstop without respect for human sleep. When large sums of money are at stake, large gains or losses occur when opportunities are met or ignored.

Even at the local level, technology makes sure that the manager or engineer never gets away. In the old days, voice-only telephones were the means of contact, placing the manager at some distance when off duty. Now managers can have computer terminals in their bedrooms, if they so choose, showing the vital process control functions of the plant or refinery, enabling them to problem-solve at any hour, even when at home.

Similarly, the fax machine and the worldwide network of telephone lines keeps executives, attorneys, and professionals constantly in touch, even when they would rather be incommunicado. A colleague makes a trip to France or Singapore, and she cannot escape. She cannot blame the postman for a delay in delivery—the sender knows when and where the missive has arrived. She has no excuse to take a break or escape—her boss expects her always to be available, to do her job, wherever she may be.

TEACHING TECHNOLOGY TO HAVE RESPECT

HOW CAN WE cope with this strange world? How can we maintain our sense of equilibrium in the face of ceaseless demand and opportunity? Clearly, we must rethink our ways of doing things and construct a world in which we each can flourish. We cannot let technology control our lives according to its own whims. We have to construct a human-centered world that knows when to draw the line. We have to take back charge over the technology and let it know who is boss. We need to "teach" it respect for human rates of data assimilation; respect for human schedules of sleep, wake, work, and relaxation; respect for human-centered environments. We cannot let it operate at its own convenience.

How do we teach an inanimate object, a technology, to pay due respect? By changing our own mind-set. We—the engineers, designers, purchasers, installers, and users of the technology— must demand machines that respect their subservient role to human needs.

TAKING CHARGE OF OUR TEMPORAL NICHE

WE CAN START by defining our personal schedule to determine in which hours we obtain the most recuperative sleep, in which hours we are most creative and best at solving problems, and in which hours we most enjoy relaxation. These need not be single blocks of time in a twenty-four-hour day. There may be more than one time to sleep—siesta hour, for example.

Next, what are the personal schedule needs of the significant others who share your personal life—your family, roommate, spouse, and children? Compromises need to be made.

Carefully analyze your job. Determine the key times that cannot be changed when communication with others must be accomplished or the needs of business must be addressed. But do not take the status quo for granted; machines, employees, and even bosses can change.

The aim is to create a work and relaxation pattern that takes care of business but also takes care of the soul—that keeps you at your best and puts technology in its place as a follower in your personal pack, not the leader. Such principles have never been more urgent than they are today. Before the twenty-four-hour society took hold, everyone worked and slept more or less on the same schedule. Certain times within the twenty-four hours were considered sacrosanct; for example, people did not invade each other's night hours except in a true emergency. Today we must create that sacrosanct time for ourselves. A fundamental human need, it must be made top priority.

The physiology of biological clocks tells us that there is no particular reason why we must sleep at night and work during the day—provided we take control over exposure to bright light and sleep schedules. We have the freedom—and responsibility—to choose our own schedule. But we must make a conscious decision to construct a personal space in time—a time cocoon for ourselves.

CREATING A TIME COCOON

CREATION OF a time cocoon means controlling the incoming flow of information, of demands on ourselves. We have to create a buffer between the world and our personally chosen schedule. Information and the expectation that information will be instantly acted on must be controlled. Provision must be made for true emergencies, of course, but these should be carefully limited intrusions on the cocoon.

The means to do this are available. Answering machines and voice-mail, modem connections, E-mail, and fax machines, if used appropriately, can create a time buffer between transmission of a message and when it is conveyed to the intended recipient. However, their use as human-centered devices relies as much on the sender of the information or query as on the recipient. It is vital not to confuse the rapidity of information transfer with the rapidity of human response. Technology has rapidly speeded up the former, but done little to speed the latter. People still need and should be given the time to develop and provide a carefully thought-out response.

Just because a document can be sent to Australia or Japan within a few seconds—instead of weeks by ship or days by airmail—does not mean that the work-rest schedule of the recipient should be disrupted. The storage buffer, whether it be the tape on an answering machine, the memory chip of a computer or a fax in-tray, should be used to hold the information until the recipient is in an alert state, ready to process it.

The time cocoon provides a respectful space for our human machine. It allows the continuous receipt of information but controls the bombardment of the individual. A day's work can be planned and deliberate attention be paid to what is truly important.

To write this book, I had to create such a time cocoon. I was running a research institute that was undergoing a major restructuring of its activities and a consulting firm that was expanding its business to Europe; several new research contracts and product rights agreements were under negotiation; and my wife and family had their own demanding work and school priorities. Faxes were pouring in from Japan and Europe; questions, concerns, and opportunities were flooding my day and night. As if this weren't enough, both my father and my father-in-law were admitted to the hospital on opposite sides of the globe.

My solution was to develop a schedule suited to my biological clock and sleep propensity that still allowed me to perform my executive and management duties and my family responsibilities. For six weeks I worked two shifts. The first, from 4:00 A.M. to noon, was protected time to write in my home office, insulated by my staff and family from the flow of information and requests. I chose this time because of my "lark" tendencies, and I could sustain it if I went to bed by 10:00 P.M. In the afternoon I went to the office and responded to the in-tray, took phone calls, and tackled the challenges of the day. From 1:00 to 5:00 P.M., I opened my window to the world. The evenings were kept clear for family matters and for relaxation.

With such time cocoon arrangements, it should not matter to the sender—except in emergencies—what time of day or night the recipient of a query or information works on his or her in-tray. We need to push back the boundaries of expectation to the time frame of the prefax era. It is reasonable to expect a response on a brief, important (but nonemergency) matter within twenty-four hours, allowing time for the recipient to follow his or her personal work schedule of choice. To expect a response in less time may be infringing on the person's boundaries.

A business etiquette that protects the rights of the citizens of our newly constituted global village needs to be established. Among the rules proscribed by the "Emily Post" of the twenty-first century should be the directive that senders of inquiries should bend their schedule to that of recipients if they desire a response in less than twenty-four hours. The etiquette of this new era will also require individuals to make known their hours of availability to process urgent requests and their status in terms of ability to act. Electronic searches and networks can easily keep track of this information and make it directly accessible on a selective basis to legitimate electronic or automated telephonic enquiries.

Sometimes direct interaction will be required via teleconference or travel. Brainstorming and creative problem solving may in some cases be much more effective when the parties are face to face in reality or electronically. In these cases, schedules must be synchronized, still respecting each individual's time cocoon and minimizing disruption of sleep and wake, work and relaxation.

The technology of videoconferencing provides its own special challenges. Video communication is rapidly reaching the point where it will greatly influence the decision whether to travel. As it becomes a standard option for every telephone, new rules of etiquette and restraint must be developed. For example, I may be working at home, but if we are electronically interconnected you don't have to know where I am. You don't need to know whether I am still in my pajamas or have just taken a shower. The videophone, however, once it becomes cheap and universal, will create expectations that it be used, and it could become an intrusion on an individual's privacy.

Radio and cable television now provide a constant source of news and entertainment on a bewildering variety of channels around the clock. Even though these information sources are fully optional, they do not provide time-independent service for the most part. Most major networks offer news only in narrow time slots. If you arrive home after 7:30 P.M. or go to bed before 11:00 P.M. in the eastern United States, the network TV news is not readily available. Fortunately, with a VCR, one can automatically record a show and view it at a self-chosen time—a task made easier by the programming codes and VCR programming boxes now becoming available.

The exception, of course, is CNN, an example of technology being applied to the realities of the global village. Ted Turner's genius was seeing the world as it is, of creating a common experience and information source for world leaders, political and business, and for citizens and travelers everywhere. The ultimate in human-centered news is the CNN Headline News repeating endlessly around the clock; whatever time of day it is, whatever time zone or virtually whatever country, people can gather the information on their own personally chosen schedule.

This is the knitting that will bind together our global village. We can take advantage of technological capabilities and at the same time also preserve human capabilities.

THE ETIQUETTE OF TRAVEL

UNLIKE SOME, I don't believe that videoconferencing will eventually eliminate business travel. There is real value to direct human contact on carefully selected occasions, partly because one can be fooled too easily by video images. Just a small stage set can convince video watchers of an impressive operation.

Furthermore, our electronic hookups deprive us of human chemistry, of friendships that through joint activities might develop, of using our multiple senses. I believe that business travel is here to stay for the foreseeable future. Although it may be less needed in some interactions because of the important role that videoconferencing will play, the expansion of worldwide trade and enterprise will more than compensate.

Given that business travel must inevitably expand, special attention must be paid to optimizing its value. Chapter 14 described new technologies that are emerging to reset the biological clock using bright light and specially timed treatments. These ease the transition between time zones and speed up the adjustment to new schedules of work and sleep. However, they do not take care of a traveler's every need. An etiquette must be developed that protects travelers in terms of travel schedules, accommodations, and meeting schedules. Flexibility should be shown by hosts to ensure the smooth adaptation of the traveler and allow time for personal schedules to be observed.

Airline travel schedules are designed too much for the economic convenience of the airlines, unnecessarily depriving the traveler of sleep because only night flights are offered on certain routes. Then, at many hotels, maids knock repeatedly on the door, wanting to clean the room, without regard for the traveler's sleep-wake schedule. Cocktail hours are held only paper-thin walls away from the traveler trying to sleep at an unconventional hour. The blockage of bright sunlight from penetrating the room is rarely achievable because of the flimsiness of drapes. Checkout times show no respect for the traveler's time cocoon.

The time cocoon principle is never more important than when one is traveling to a distant time zone. Human alertness technology needs to reach the hotel industry. Despite all we hear about low occupancy rates and cutthroat competition, most hotels have yet to address the true needs of the international traveler.

To correct this situation, Circadian Technologies has created a *"Circadian Standard™"* for the hotel industry that will provide the international traveler with a critically needed time cocoon. To meet this standard, a hotel must provide executive-quality rooms that are rigorously equipped and serviced to provide the opportunity, for sleep, meals, work, and relaxation to the traveler on any time schedule. Blackout drapes allow complete control of light and dark, rooms are placed on quiet corridors away from traffic and hotel noise, meals are available twenty-four hours a day, and maid and other hotel services are sensitive to the traveler's schedule. Furthermore, human alertness technologies to combat jet lag are available to the guest.

Hotels in the major cities of the world are already making a commitment to the Circadian Standard. At last, travelers have a chance to find the support they need to maintain a time cocoon wherever in the world they may be.

Chapter **19**

Preparing for the Twenty-First Century

NOW, ON THE BRINK of entering a new century, is the time to think hard about our "New Century's Resolutions." It is time to take stock of where we have come in the twentieth century and where we are going in the next. We need to decide now whether we are on the right express train, for it is accelerating at a frightening pace.

The technological changes in this past century have been remarkable. At the beginning of this century, electric light was barely invented, we had no airplanes, no computers, no spaceships, no TV. Automobiles were primitive, homes were devoid of electronic gadgets, warfare was without nuclear weaponry, and travel by land or sea was conducted at a leisurely speed. Crossing continents or oceans took many days, and the telephone was a new invention just beginning to find its place.

Engineers have ruled supreme this century. They have given us a mastery over our environment and have converted machines from temperamental gadgets to reliable, nearly fail-safe devices. They have miniaturized and microprocessorized our world, and with so much power at our fingertips we can satisfy virtually every whim. We can decide to take a man to the moon, or to Mars, and the only question is whether we have the budget to do it. Never has the "can-do" attitude been so alive; never has technological innovation created so many technologies and gadgets anticipating wishes we never knew we would have.

But within our technological world lie the seeds of its own destruction. The red lights are flashing, the warning bells are clanging, the storm clouds approaching—for our technology has forgotten the needs of the human race from which it was spawned.

By failing to understand the physiological needs of our bodies, by ignoring human limitations, we have followed the dictates of technology and economics into a world that risks becoming unfit for the human race. We have tied together nations into one big global village but forgotten the temporal dimension that lies at the essence of our being.

If we don't mend our ways and overhaul our technology to build a human-centered world, we will find ourselves cast out from our home of origin with nowhere to go. Technology is a wonderful servant but a terrible master, and yet we are grooming it for that ultimate promotion.

Unless we act now, we will have more Chernobyls, Three Mile Islands, Bhopals, and *Exxon Valdez*es on a larger and larger scale. We will have more and more airplane accidents, highway crashes, and plant explosions due to human fatigue. We will steadily increase the ill health of our citizens as we increase the disparity between human wants and human needs. We may want to achieve foolish dreams that ignore the physiology of our bodies, but we will only suffer from the creation of such incompatibility.

WISDOM WILL REWARD YOU

AS THE BIBLE verse says, "If you are wise, your wisdom will reward you." The wise man on seeing the coming storm gathers his flocks into his shelter. But we have the chance to do much more than that. For this storm is of our own making. We do not have to suffer its consequences. We must see beyond what is familiar and refuse to accept placidly paradigms in urgent need of change.

Ten years ago, when I first realized the magnitude of the problems we face, I mistakenly thought that just telling people with the power to act would be all that was needed. After all, the problems and the solutions seem so obvious, although often challenging to implement.

My colleagues and I have testified before Congress, published articles, and given thousands of speeches to corporate leaders, managers, and executives across North America and Europe. We have given TV, newspaper and magazine interviews, and spoken with the engineers that build the devices that cause people the most trouble.

Certainly, we found champions and enthusiasts, and much has been accomplished on a limited scale. Thousands of plants changed their shift schedules, started to train their employees and managers, and this is now all happening at an ever-expanding pace. But the needed move to a human-centered society is stymied by the inertia of key members of our society—especially the engineers, who have not moved beyond their old way of thinking. Machine-centered thinking, when it permeates all our industries, our technology and our equipment, imprisons our society in dangerous paradigms of thought.

THINK GLOBALLY, ACT LOCALLY

NEVER HAS so much wisdom been conveyed by a bumper sticker. The world becomes year by year a smaller place. We are increasingly becoming citizens of a planet rather than of individual nations.

All of us must therefore take individual responsibility for changing the world in which we live. Not only the engineers, corporate chieftains, and politicians have to change their mindsets. We set their design specs, buy their products, and elect them to office.

As individual citizens, we need to ask our leaders and politicians what is being done to reduce the risk of errors caused by human fatigue. As consumers, we need to demand the services that enable us to protect our own time cocoons. As purchasers, we need to look for human alertness technologies to enhance our world. Even relatively small groups of vocal and active concerned citizens can do much to shape ways of thinking and decision making.

Ask airlines about the schedules the crews are asked to fly; what steps are they taking to abolish fatigue? Ask hospitals how many consecutive hours the resident physicians and surgeons are kept on duty. Ask bus and railroad companies how long their drivers or engineers are required to stay awake. It is your own personal safety at stake.

As free-market economies span the globe, as the pool of entre-preneurial talent expands on all continents of the planet, we have the capacity to meet the needs of the human race—if we know in which direction to head. As both citizens and leaders become informed and start to buy the human-centered technology, free enterprise will oblige with the supply.

There is much richness to be gained by life in the global village, but we must make sure that it is fit for human habitation. In the twentieth century, we made ourselves technological giants. To succeed in the twenty-first century, we need to gain more wisdom about ourselves. In making our new century's resolutions, we must heed the old truth with new meaning:

You made my body, Lord; now give me sense to heed your laws.

Psalms 119:73

Source Notes

Foreword to the British Edition

Page xvii   British Airways BAC 1–11 incident: Air Accidents Investigation Branch Report 1/92, Department of Transport, April 1992; *Daily Telegraph*, April 10, 1992, and conversation with British Airways safety personnel.

chapter 1   Outpaced By Our Technology

Page 3   O'Hare incident: Report submitted to Air Line Pilots Association (ALPA), May 1983.

Page 3   *Vincennes* incident: *Business Week,* September 12, 1988, p. 36, and from conversations with members of U.S. Navy investigation team.

Page 4   A bomb, planted by terrorists in a passenger's luggage on Pan Am flight 103 from London to New York, exploded over Lockerbie, Scotland, on December 21, 1988, causing the disintegration and crash of the plane, and killing 259 passengers and crew plus 11 people on the ground.

Page 5   Major nighttime accidents have included Three Mile Island (4:00 A.M., March 28, 1979); Bhopal (12:15 A.M., December 3, 1984); Chernobyl (1:23 A.M., April 26, 1986); and *Exxon Valdez* (12:04 A.M., March 24, 1989).

Page 6   The role of human fatigue in the fateful decision to launch the NASA space shuttle *Challenger* flight 51-L, on January 28, 1986, is documented in Appendix G, "Human Factors Analysis," of the *Report of the Presidential Commission on the Space Shuttle* Challenger *Accident.* This report also documents the considerable problems of fatigue and fatigued-induced errors in the technical teams working long hours preparing the shuttle for launch.

Page 7   Medium- and heavy-truck accidents involving fatalities cost on average $1.5 million, according to U.S. Department of Transportation estimates.

The estimates of Exxon's legal liabilities from the *Valdez* incident come from discussions with various well-informed sources within the petrochemical industry.

Unocal's tape is an internal production for informing employees of these risks.

Page 7   I first coined the term "twenty-four-hour society" in my testi-

215

mony on "Biological Clocks and Shift Work Scheduling" before
the Subcommittee on Investigations and Oversight of the House
Committee on Science and Technology, chaired by Congressman
Al Gore, March 23, 1983.

Page 8   Numerous sources discuss the agricultural and industrial revolu-
            tions. A succinct account can be found in *The New Columbia
            Encyclopedia* (Columbia University Press) or other equivalent
            sources.

Page 10   "The mountains are high and the emperor is far away": see
            Forbes, August 5, 1991. Also see "A Chinese Village Discovers
            the Road to Riches," *New York Times,* January 10, 1992.

chapter **2**   **Facing Up to the Challenge**

Page 11   The extended hours on duty of my internship (first-year resi-
            dent, house officer) experience in 1970–1971 in Toronto, Can-
            ada, were no different from those experienced by newly
            graduated physicians in many countries, even today. See Chap-
            ter 7 for a fuller discussion.

Page 12   The biological clock (supra chiasmatic nucleus) in the mamma-
            lian brain that generates circadian rhythms was not positively
            identified until 1972, by two independent groups of investiga-
            tors: F. K. Stephan and I. Zucker. Proceedings of the National
            Academy of Sciences (USA) 69:1583–1586 (1972); R. Y. Moore
            and V. B. Eichler, *Brain* Research 42:201–206 (1972). Also see
            M. C. Moore-Ede, F. M. Sulzman, and C. A. Fuller, *The Clocks
            That Time Us* (Harvard University Press, 1982), and Chapter 3
            for fuller discussion.

Page 12   We set up the first human time isolation facility in the United
            States as a collaborative venture using the top floor of the
            Department of Neurology at the Montefiore Hospital in New
            York. Eliot Weitzman was professor of neurology at Albert Ein-
            stein and chairman of neurology at Montefiore Hospital. Charles
            Czeisler, who had done his undergraduate thesis at Harvard Col-
            lege under my supervision, was a Stanford University medical
            student at the time. He is now associate professor of medicine at
            Brigham and Women's Hospital, Boston. Richard Kronauer, Gor-
            don Mackay Professor of Mechanical Engineering at Harvard, is
            an applied mathematician who has had considerable experience
            analyzing oscillating systems, and contributed significantly to
            the theory of circadian clocks.

            Research on human circadian rhythms was pioneered by Profes-
            sors Jurgen Aschoff and Rutger Wever at the Max Planck Insti-
            tute fur Verhaltensphysiologie at Erling-Andechs using a
            converted underground bunker. Located in a beautiful site south
            of Munich, near the famous monastery in the foothills of the
            Alps, Erling Andechs served as an intellectual home for the
            newly developing field. For a more complete description of the

history and an introduction to a comprehensive literature, see M. C. Moore-Ede et al., *The Clocks That Time Us.*

Page 12   To help with the project, we brought in at a later stage a Stanford University psychologist, Richard Coleman.

Page 13   The scientific report of this first application of human circadian physiology to industrial shift schedules was published in C. A. Czeisler, M. C. Moore-Ede, and R. M. Coleman, *Science* 217:460–463 (1982).

Page 17   Corporate sponsors who have provided vital and much-appreciated support of the institute's research programs include Alcan Aluminum, Alexander & Alexander, Amoco, Atlantic Richfield, Boeing, Boston Edison, Cabot, Caterpillar, Conrail, Dow Chemical, DuPont, Exxon, Foxboro, Liberty Mutual, Madison Gas & Electric, Matsushita Electric Works, Mobil, Monsanto, Raytheon, Southern California Edison, Sun Refining and Marketing Company, Texaco, and 3M.

chapter **3**   **Design Specs of the Human Machine**

Page 19   I worked as a research associate at the NASA Ames Research Center with Professor John West during the summer of 1967 undertaking computer modeling of the distribution of blood and gas in the lung during weightlessness, and hence had a brief opportunity to be a part of the early NASA days before the Apollo space program first landed men on the moon in July 1969. Since 1974 I have been a principal investigator on a number of spaceflight experiments and other NASA contracts.

Page 22   The brief discussion of the physiology of biological clocks and sleep-wake cycles in this chapter is much more extensively covered in M. C. Moore-Ede et al., *The Clocks That Time Us,* which also contains a comprehensive listing of references in the scientific literature. For a readable and informative discussion of the physiology of sleep for the general reader, W. C. Dement, *Some Must Watch While Some Must Sleep* (W. H. Freeman, 1974), is an excellent source.

Page 26   Figure 3.5 is from P. Hauri, *The Sleep Disorders* (The Upjohn Company, 1982), with permission.

Page 32   The role the biological clock plays in predicting the future and the balance that must be achieved between predictive and reactive homeostasis is discussed in my Bowditch Lecture to the American Physiological Society; see M. C. Moore-Ede, *American Journal of Physiology* 19:R735–R752 (1985).

Page 33   Walter B. Cannon, professor of physiology at Harvard Medical School, is recognized as the father of American physiology. He developed the concept of homeostasis between 1910 and 1932.

Page 34   The first human time isolation experiments were done in a

French cave by Michel Siffre, but the first comprehensive exploration was done by Aschoff and Wever, as noted in Chapter 2 notes above. Also see R. A. Wever, *The Circadian System of Man* (Springer-Verlag, 1979).

Page 37 The relationship between bedtime and the length of sleep: C. A. Czeisler, E. D. Weitzman, M. C. Moore-Ede et al., *Science* 210:1264–1267 (1980).

Page 38 Human circadian rhythm range of entrainment: R. A. Wever, *The Circadian System of Man.*

Page 39 Figure 3.10 was redrawn from J. R. Beljan et al., *Human Performance in the Aviation Environment,* NASA Contract 2-6657, 1972.

Page 40 Light-dark cycles as human time cue: C. A. Czeisler, G. S. Richardson, J. C. Zimmerman, M. C. Moore-Ede, and E. D. Weitzman, *Photochemistry and Photobiology* 34:239–247 (1981).

Page 41 Phase response curve to light: For historical review and detailed discussion of the mechanisms, see M. C. Moore-Ede et al., *The Clocks That Time Us*; for human phase response curve see R. A. Wever, *Journal of Biological Rhythms* 4:161–185 (1989); C. A. Czeisler et al., *Science* 244:1328–1332 (1989); D. Dawson et al., *Sleep Research* 18:413 (1989).

Page 42 The Samuel Johnson quote is from Boswell's *Tour of the Hebrides,* September 14, 1773.

chapter 4 Alertness: The Achilles Heel of a Nonstop World

Page 47 The experimental psychology literature uses the term *arousal* for the state of activation of the brain, whereas the sleep physiology literature uses the term *alertness.* I prefer *alertness* because of the unfortunate connotations of the term *arousal.* Consider "the pilot of my plane was aroused" as compared to "the pilot of my plane was alert"! Which makes you feel safer?

Page 49 The anatomy and physiology of the sympathetic and parasympathetic nervous systems was defined by Walter B. Cannon in his *Wisdom of the Body* (W. W. Norton, 1932) and is explained in any textbook on mammalian physiology.

Page 51 For a discussion of sleep stages, see Dement's *Some Must Watch While Some Must Sleep.*

Page 51 The Multiple Sleep Latency Test is discussed in various sources, including M. A. Carskadon and W. C. Dement, *Neuroscience and Biobehavioral Reviews* 11:307–317 (1987); G. S. Richardson et al., *Sleep* 5:S82–94 (1982); J. K. Walsh et al., *Sleep* 11:251–264 (1988). MSLT functions are derived from these sources and data from the Institute for Circadian Physiology.

chapter **5**   **The Costs of Human Breakdown**

Page 64   *Exxon Valdez* incident: *New York Times,* November 19, 1989, March 24, 1990, August 1, 1990, April 10, 1991.

Page 66   Chernobyl incident: Zhores A. Medvedev, *The Legacy of Chernobyl* (W. W. Norton, 1990).

Page 67   Total U.S. truck accident costs: 1988 NTSA NASS/GES estimates for medium- and heavy-truck accidents: 4,893 fatal accidents at $1,500,000 average each; 45,433 injury accidents at $19,000 average each; 131,168 property damage accidents at $6,000 average each. Estimated total $8.989 billion.

Proportion of truck accidents where driver fatigue is a major factor: National Transportation Safety Board, *Safety Study— Fatigue, Alcohol, Other Drugs, and Medical Factors in Fatal to the Driver Heavy Truck Crashes,* vol. 1 NTSB/SS-90/01 (National Technical Information Service, 1990).

Cost of major airline accidents: J. Lauber and P. Kayten, keynote address to second annual meeting of Association of Professional Sleep Societies, June 12, 1988.

Proportion of aviation accidents due to human error: The National Plan for Aviation Human Factors, Federal Aviation Administration, December 1990.

Page 67   Data on increased rate of shiftworker highway accidents: Circadian Technologies confidential surveys of a total of more than 12,000 shiftworkers and dayworkers in industrial plants in North America.

Annual cost of highway accidents: National Safety Council as cited by *World Almanac* (1987), p. 772. Also annual cost of industrial accidents from same source.

Page 69   Value added by manufacturing: *World Almanac* (1987), p. 134.

Peach Bottom costs: *New York Times,* March 27, 1988. Also see Chapter 8.

Page 71   Shift maladaptation syndrome: M. C. Moore-Ede and G. S. Richardson, *Annual Reviews of Medicine* 36:607–617 (1985), also contains references to studies on the medical consequences of shift work.

Page 76   Total cost of coronary artery disease: calculated from data indicating total cost of cardiovascular disease is $85 billion per year and heart attacks represent 55 percent of these cases. Source is National Center for Health Statistics and American Heart Association as cited by *World Almanac* (1987) p. 87.

Page 77   Two shiftworker families with children: See, for example, *Boston Globe,* September 26, 1989.

Page 77   Around-the-clock child care centers: *Arizona Republic,* March 18, 1990.

Increase in divorce rate: Studies by Professor Donald Tepas, University of Connecticut, Storrs, personal communication.

Page 78　Quote from shift worker: *The Living Clock,* PBS television documentary in Infinite Voyage Series, WQED Pittsburg.

chapter **6**　**Aviation Safety and Pilot Error**

Page 81　Lockheed Tristar incident: CHIRP reporting system, RAF Institute of Aviation Medicine, Farnborough, Hants.

Page 81　Proportion of aviation accidents involving human error: The National Plan for Aviation Human Factors, Federal Aviation Administration, December 1990.

Page 82　Cost of major airline accidents: J. Lauber and P. Kayten, keynote address to second annual meeting of Association of Professional Sleep Societies, June 12, 1988.

FAA denial: For example, see *New York Times,* June 3, 1988.

Second officer report on multiple fatigue-related incidents: Report submitted to Air Line Pilots Association (ALPA), May 1983.

Page 83　Boeing 707 incident: *U.S. Army Digest* 24:12–13 (1978).

Page 83　Aerospace Medical Association annual meeting 1990.

Page 83　Unpublished data from Circadian Technologies study of aircrew alertness and sleep patterns.

Page 85　Pan Am Boeing 707 accident in Bali: Discussed in M. C. Moore-Ede et al., *The Clocks That Time Us.*

Page 86　China Airlines incident: *New York Times Magazine,* March 27, 1988.

Page 86　Other fatigue-related aviation accidents: W. J. Price and D. C. Holley, *Occupational Medicine: State of Art Reviews* 5:343–377 (1990).

Page 88　British Airways BAC 1–11, see Foreword.

Page 90　Effects of automation on pilot skills and performance: C. E. Billings, *Challenges in Aviation Human Factors: The National Plan,* American Institute of Aeronautics and Astronautics Conference: Vienna, Virginia, 15–16 January 1991, Abstracts, pp. 3–5.

Page 90　AERA-2: Ibid., pp. 76–77, 123–126; also see J. C. Celio, *Controller Perspective of AERA-2,* Mitre Corp., MP-88W00015 (1990).

Page 91　The National Plan for Aviation Human Factors, Federal Aviation Administration, December 1990.

Page 92　Studies on pilot napping: R. C. Graber, *Challenges in Aviation Human Factors* (1991); M. R. Rosekind et al., *Sleep Research* 20 (1991); L. J. Connell, *Sleep Research* 20 (1991).

Page 92   British Aircrew Regulations: The Avoidance of Fatigue in
           Aircrew: Guide to Requirements (CAP 371) Third Edition, May
           1, 1990, Civil Aviation Authority.

chapter **7**   **Medical Care Whenever You Need It**

Page 98   Californian incidents of sleepy residents' errors: *Time,* December
           17, 1990.

Page 99   Resident quote: U.S. Congress, Office of Technology Assessment,
           Biological Rhythms: Implications for the Worker, OTA-BA-463,
           Washington, D.C., September 1991, p. 166 (OTA Report). Nurse
           quote: personal interview.

Page 100   Virginia Prego case: *New York Times,* February 15, 1990.

Page 100   Resident made quadriplegic: OTA Report, p. 166.

Page 102   Libby Zion case: *New York Times,* January 13, 1987.

           Axelrod initiative: *New York Times,* May 30, 1987; also *Time,*
           August 31, 1987; also OTA Report, pp. 168–170, provides
           detailed discussion of the regulatory changes since the Zion
           incident.

Page 104   Bloomsbury Health Authority case: OTA Report, p. 171.

Page 104   New Zealand resident hours of work: OTA Report, p. 172.

chapter **8**   **The Power of Our Society**

Page 108   Zhores A. Medvedev, *The Legacy of Chernobyl.*

Page 109   UK radiation fall-out from Chernobyl: National Radiological
           Protection Board as reported in *Daily Telegraph* May 5, 1986.

Page 110   Peach Bottom: *New York Times,* March 27, 1988, August 12,
           1988; *Wall Street Journal,* April 1, 1987, November 5, 1990.

Page 111   See Circadian Technologies, "Control Room Operator Alertness
           and Performance in Nuclear Power Plants," prepared for Elec-
           tric Power Research Institute (EPRI Report No. NP-6748, proj-
           ect 2184-7) (1990).

chapter **9**   **Keep On Trucking**

Page 117   M42 Traffic Accident: *The Independent* June 17, 1992, and inter-
           view with expert witness at trial.

Page 117   Gloria Estefan case: *Quill Newsletter of Loss Prevention,* vol. 5,
           no. 1 (1992), and from interview with Richard Russo, attorney to
           Estefan.

Page 119   Oklahoma Turnpike Authority: from H. W. Case and S. Hulbert,
           Report 70-60 UCLA Dept. Engineering (1970).

Page 119   California Division of Highways: from J. F. O'Hanlon and G. R. Kelley, Human Factors Research, Inc., Technical Report 1736-F.

Page 119   See Chapter 5 for source of accident cost estimates.

Page 120   Claudio Stampi, personal communication.

Figure 9.1 plotted from Mitler, M. M. et al., *Sleep* 11:100–109 (1988), with permission.

Page 121   Truck driver report from videotape *Staying Awake, Staying Alive* (MEGA Safe Training Company, 1989).

Page 125   Clapham Junction Accident: Investigation into the Clapham Junction Railway Accident, Anthony Hidden QC, Department of Transport, November 1989; *Daily Telegraph* December 13, 1988; ibid March 16, 1989.

Page 127   Burlington Northern accidents: R. M. Coleman, *Wide Awake at 3 A.M.* (W. H. Freeman, 1986).

Page 127   Accident with engineer awake for twenty-six hours: *The Quill*, vol. 4, no. 11 (1991).

Page 127   Railroad engineer EEG recording: L. Torsvall and T. Akerstedt, *Electroencephalography Clinical Neurophysiology* 66:502–511 (1987).

Page 129   Zeebrugge accident: mv *Herald of Free Enterprise* Report of Court 8074 Formal Investigation, Department of Transport, July 24, 1987.

chapter 10   Decision-Maker Fatigue

Page 130   Space shuttle *Challenger* launch decision: Appendix G, "Human Factors Analysis," of *Report of the Presidential Commission on the Space Shuttle* Challenger *Accident*.

Page 133   Loss of creative thinking with fatigue: J. A. Horne, *Sleep* 11:528–536 (1988). Also see *New York Times*, January 5, 1989.

Page 135   Haig shuttle diplomacy: *Time*, April 26, 1982.

chapter 11   Human Fatigue and the Law

Page 138   Matthew Theurer case: *Wall Street Journal*, April 16, 1991; *New York Times*, April 1, 1991.

Page 139   *Exxon Valdez:* Sources cited under Chapter 5 notes.

Page 139   Peach Bottom case: Sources cited under Chapter 8 notes.

Page 140   Bhopal indictment and extradition: *New York Times*, April 15, 1992.

Page 141   Measurement of alertness: See Chapter 4 notes.

Page 141   Tugboat case: *McAllister Brothers, Inc. et al.* v. *TSS Festivale et al.*, United States District Court for the District of Puerto Rico.

Page 146   Truck accident case: *Sheryl Wyatt* v. *Unichem International et al.,* New Mexico District Court.

Page 147   Effect of short naps: C. S. Stampi, *Why We Nap: Evolution, Chronobiology, and Function of Polyphasic and Ultrashort Sleep* (Birkhäuser Boston, 1992); C. S. Stampi, *Sleep Research* 20:471 (1991); *Sleep Research* 19:408 (1991).

chapter 12   Monitoring the Status of the Human Brain

Page 154   Eastern Airline crash: see *New York Times Magazine,* March 27, 1988.

Page 155   NASA pilot fatigue studies: see Chapter 6 notes.

chapter 13   Boxing Up the Sun

Page 161   Bright light needed to suppress melatonin in humans: A. Lewy et al., *Science* 210:1267–1269 (1980).

Page 162   Seasonal affective disorder: see M. C. Blehar and A. J. Lewy, *Psychopharmacology Bulletin* 26:465–494 (1990); N. E. Rosenthal et al., *Archives of General Psychiatry* 41:72–80 (1984); J. S. Terman et al., *Neuropsychopharmacology* 2:1–22 (1989); M. Terman et al., *Psychopharmacology Bulletin* 26:505–510 (1990).

Page 163   Light effect on night-shift alertness: S. S. Campbell and W. Dawson, *Physiology and Behavior* 48:317–320 (1990).

Page 164   Dawn-dusk simulations: M. Terman, personal communication.

Page 165   Light visor: M. I. Edelson et al., Sleep Research 1991; Levitt et al., 1991. G. Brainerd, personal communication.

chapter 14   Traveling Through Time

Page 169   Phase response curve to light: see Chapter 3 notes.

Page 169   MidnightSun: T. A. Houpt, Z. B. Boulos and M. C. Moore-Ede, presented at annual meeting of Society for Biological Rhythms, May 1992.

Page 171   Melatonin clock-resetting: J. Arendt et al., *Ergonomics* 30:1379–1393 (1987); A. Lewy et al., *Advances in Pineal Research* 5:285–293 (1991).

Page 171   For brief review of pharmacology of clock-resetting, see OTA Report, pp 56–57.

chapter **15** **The Human-Centered Workplace**

Page 177   Napping strategies: C. S. Stampi, see Chapter 11 notes.

Page 178   Army study: Reported by Belinsky at Office of Technology
            Assessment Workshop on Shift Work and Extended Duty Hours,
            May 23, 1990.

chapter **16** **National Challenges in a Nonstop World**

Page 188   A study of twelve-hour shifts has recently been completed by the
            Institute for Circadian Physiology for the Nuclear Regulatory
            Commission showing little evidence for any deterioration of
            human alertness or performance on twelve-hour shifts as com-
            pared to eight-hour shifts.

chapter **18** **Developing Respect for the Human Machine**

Page 198   President Bush incident: *New York Times,* January 9, 1992;
            Time, January 20, 1992.

Page 201   Questionnaire to detect morningness-eveningness types: J. A.
            Horne and O. Ostberg, *International Journal of Chronobiology*
            4:97–110 (1976).

Index

Accidents
  airline, 3, 67, 68, 81–88, 154–155, 219n, 220n
  cost of, 64–78
  highway, 28, 67, 68, 100, 117–121, 138, 146–148, 155, 219n
  industrial, 15, 66–67, 68
  litigation related to, 138–148
  nuclear power plant, 5–6, 66, 108–110
  railroad, 125–128
Agricultural revolution, 8, 9
Airline accidents
  attentiveness failure and, 154–155
  cost of, 67, 68, 219n, 220n
  fatigue-related, 3, 81–88
Airline safety
  automation as hazard to, 89–92
  cockpit design in, 89, 91
  controlled napping and, 92–93
  fatigue management and, 95–96
  flight-time/duty-time regulations and, 92, 93–95
  goals for, 89
Air traffic controllers, 88, 90–91
Air travel
  alertness technology in, 168–72
  biological clock and, 38–39, 41–42, 135, 166–168, 198, 208
  hotel accommodations and, 208–209
  schedules, 208
  vs telecommunications, 172, 207–208
Akerstedt, Tjorborn, 127
Alcan, 202
Alertness
  assessing impairment to, 142–146
  vs comfort, 62–63
  component of attentiveness, 47–48
  ingredients of, 49–50
  levels, 28
  management of, 48–49, 114, 199–200
  measuring, 50–53
  postlunch dip, 27–28, 120, 174
  switches triggering, 53–62, 123
  See also Biological clock

Alertness/attentiveness monitoring, 151–152
  alertness monitoring, 154–156
  attentiveness measurement, 153–154
  computer performance tests, 152–153
  feedback system in, 156–157
  job definition and, 157–158
  multiple sensory cues and, 158–159
Alertness technology, 181–182
  biological clock resetting, 168–172
  during conference room presentations, 173–176
  in emergency situations, 176–179
  in global village, 201–203, 207
  light levels and, 162–165
  personal schedule development and, 203–207
  recuperation facilities, 179–180
  sleep quality and, 180–181
Alpha waves, 51
American Airlines, 3
Andechs, Erling, 216n
Anderson, Warren, 140
Animal
  light response, 161
  sleep, 22, 31, 36, 177
Armstrong, Neil, 14
Army, U.S., productivity study of, 178–179
Aroma, and alertness, 61–62, 176
Arousal. See Alertness
Aschoff, Jurgen, 216n
Attentiveness
  alertness and, 47–48
  traditional approach to, 46–47
  See also Alertness/attentiveness monitoring
Austin, John H., Jr., 139
Automatic behavior state, 3
Automation, 9, 89–92, 115
Automobile accidents. See Highway accidents
Autonomic nervous systems, 49–50, 157

225

For further information, please contact:

Circadian Technologies, Inc.
One Alewife Center
Cambridge
MA 02140–2317
USA

Tel: 0101 617 492 5060
Fax: 0101 617 492 1442